"This book is packed with informed and useful information. If you need to know about how sound works, and read tons of information about how classical music is recorded in the real world, then you must read this book. Perfect for anyone interested in classical music."

—Paul Baily—classical recording and post production, Re:Sound

Recording
Classical
Music

Recording Classical Music presents the fundamental principles of digitally recording and editing acoustic music in ambient spaces, focusing on stereo microphone techniques that will help musicians understand how to translate "live" environments into recorded sound.

The book covers theory and the technical aspects of recording from sound source to delivery: the nature of soundwaves and their behavior in rooms, microphone types and the techniques of recording in stereo, proximity and phase, file types, tracking and critical listening, loudness, meters, and the post-production processes of EQ, control of dynamic range (compressors, limiters, dynamic EQ, de-essers), and reverberation (both digital reflection simulation and convolution), with some discussion of commercially available digital plugins. The final part of the book applies this knowledge to common recording situations, showcasing not only strategies for recording soloists and small ensembles, along with case studies of several recordings, but also studio techniques that can enhance or replace the capture of performances in ambient spaces, such as close miking and the addition of artificial reverberation.

Recording Classical Music provides the tools necessary for anyone interested in classical music production to track, mix, and deliver audio recordings themselves or to supervise the work of others.

Robert Toft teaches in the Don Wright Faculty of Music at Western University, Canada. His interests revolve around the notion of research informing practice, and he specializes in both recording practices and the history of singing. He has written five books on historically informed vocal performance and has given master classes at leading conservatories and universities in Australia, Austria, Britain, Canada, Germany, Ireland, Switzerland, and the USA. Robert's production company, Talbot Records, released its first recording in 2017. Inspired by the intensely dramatic performing styles of the sixteenth to nineteenth centuries, its main series, Radically Hip, connects modern audiences to the impassioned eloquence of the past.

Recording
Classical
Music

Robert Toft

Routledge
Taylor & Francis Group

NEW YORK AND LONDON

First published 2020
by Routledge
52 Vanderbilt Avenue, New York, NY 10017

and by Routledge
2 Park Square, Milton Park, Abingdon, Oxon, OX14 4RN

Routledge is an imprint of the Taylor & Francis Group, an informa business

Library of Congress Cataloging-in-Publication Data
Names: Toft, Robert, author.
Title: Recording classical music / Robert Toft.
Description: New York, NY : Routledge, 2020. | Includes bibliographical references and index.
Identifiers: LCCN 2019020998 (print) | LCCN 2019022129 (ebook) | ISBN 9780815380252
    (hbk : alk. paper) | ISBN 9780815380245 (pbk : alk. paper)
Subjects: LCSH: Sound recordings—Production and direction. | Sound—Recording and
    reproducing. | Music—Acoustics and physics.
Classification: LCC ML3790 .T63 2020 (print) | LCC ML3790 (ebook) | DDC 781.68/149—dc23
LC record available at https://lccn.loc.gov/2019020998
LC ebook record available at https://lccn.loc.gov/2019022129

ISBN: 978-0-8153-8025-2 (hbk)
ISBN: 978-0-8153-8024-5 (pbk)
ISBN: 978-1-351-21378-3 (ebk)

Typeset in ITC Giovanni Std
by Apex CoVantage, LLC

Visit the companion website: www.routledge.com/cw/toft

Contents

Preface   ix
Acknowledgments   xi

PART 1 • Fundamental Principles   1
  1   Soundwaves   3
          Enclosed Spaces   7
  2   Audio Chain From Sound Source to Listener   13
          Integrity Within an Audio Chain   13
          Basic Concepts and Terminology   13

PART 2 • Production   21
  3   Microphone Types   23
          The Behavior of a Pure Diaphragm   23
          Condenser Microphones   24
          Dynamic and Ribbon Microphones   28
  4   Microphone Characteristics   31
          Frequency Response   31
          Directional (Polar) Patterns   31
          Random Energy Efficiency (REE; Also Called Directivity Factor)   33
          Distance Factor   34
          Proximity Effect   35
          Phase   37
  5   Stereo Microphone Techniques   45
          Coincident Pairs   47
          Near-Coincident Arrays   51
          Spaced Microphones   54
  6   Tracking   59
          Critical Listening   59
          Setting Levels   60
          Room Ambience   61

PART 3 • Post-Production   63
  7   EQ   65
          Digital Filters   65
          Common Practices (A Place to Start as Listening Skills Develop)   70
  8   Control of Dynamic Range   73
          Compressors   73
          Limiters   79
          Dynamic EQ   82
          De-essers   84
  9   Reverberation   89
          Digital Reflection Simulation   89
          Convolution   102

10	Delivery	105
	File Types	105
	Loudness and Meters	107
PART 4 •	Common Recording Strategies	123
11	Solo Piano	125
	Recording in Stereo	126
	Unfavorable Room Acoustics	127
12	Soloists With Piano Accompaniment	129
	Voice	129
	Violin	130
	Cello	131
	Clarinet, Oboe, Flute	131
	Trumpet, Trombone, French Horn	132
13	Small Ensembles	135
	Piano Trio	135
	String Quartet	135
	Chamber Choir	136
14	Sessions	137
	Solo Piano	137
	Solo Cello	138
	Double Bass and Piano	139
15	Studio Techniques: Re-Creating the Aural Sense of Historic Spaces	143
	Pre-Production	144
	Production	147
	Post-Production	149
Glossary		153
Index		163

Preface

This book serves as an introductory text to fundamental principles of digitally recording and editing acoustic music in ambient spaces. It focuses on stereo microphone techniques in classical genres to help performers unaccustomed to recording environments understand the processes involved in crafting records. Musicians spend thousands of hours preparing for the concert platform but relatively little time (if any) learning how to turn those performances into recorded sound. A concert hall recital, replicated in front of microphones, rarely produces a satisfactory outcome on a distribution medium such as the compact disk or mp3, for the methods engineers and producers use to shape what listeners hear through loudspeakers have an enormous impact on the way people react to recorded performances.

In fact, sound recordings have become such an important form of musical communication that all musicians probably should familiarize themselves with the procedures employed to generate the sensory surfaces of records. By providing information on the art of committing performances to disk, this book will enable musicians to turn recitals into raw tracking data that can be digitally edited into cohesive listening experiences.

The text addresses the following topics:

Theory of Sound Recording

- the nature of soundwaves and their behavior in enclosed spaces
- the components of a recording chain from sound source to listener
- the conversion of analog signals to digital information
- resolution or sound quality in digital systems

Production

- microphones
- stereo microphone techniques
- phase issues and the "three-to-one" principle
- tracking
- critical listening

Post-Production

- digital editing—EQ, control of dynamic range, reverberation, loudness, meters
- discussion of commercial software plugins
- preparing the finished track for delivery in a variety of file types (WAV, AIFF, FLAC, mp3, AAC)

Several case studies follow these theoretical sections, and they cover the basic principles behind a number of common recording situations from solo performance to various combinations of musicians and instruments—piano, soloist with piano accompaniment, and small ensembles. The book ends with an exploration of studio techniques that can enhance or replace the microphonic capture of performances in ambient rooms. In summary, the primary objective of the book is not to train musicians as recording engineers but to provide performers with a theoretical and practical understanding of how musical performance can be transferred to audio media.

The text presumes that readers are already familiar with the digital audio workstation (DAW) of their choice and know how to get sound in and out of it. Hence, instead of providing detailed discussions of specific recording software, the book introduces principles that can be translated to any DAW. The glossary contains definitions for many of the technical terms encountered in the text.

Acknowledgments

The screenshots included in the text have been used with the permission of Brainworx Audio, EastWest Communications, Exponential Audio, FabFilter Software Instruments, Flux: Sound and Picture Development, Harrison Consoles, iZotope, MeterPlugs, Nugen Audio, Sonnox, Sound Radix, and Voxengo. Suzy Lamont Photography and Queen's University (Kingston) have given permission for the use of the photograph in Figure 14.3 (the Performance Hall in the Isabel Bader Centre for the Performing Arts), and the photograph of microphone placement in one of Simon Eadon's recording sessions (Figure 14.1) has been used with Mr. Eadon's permission.

All the diagrams in the text have been drawn by the author. FabFilter Software Instruments has granted permission for the reproduction of the schematic in Figure 8.7, which depicts the action of a compressor. While every effort has been made to trace copyright holders, this has not been possible in all cases. Any omissions brought to the publisher's attention will be remedied in further editions.

PART 1

Fundamental Principles

CHAPTER 1

Soundwaves

Vibrating objects cause the compression and rarefaction of air molecules, and this alternating difference in pressure creates waves of sound (see Figure 1.1). All waveforms with pitch are periodic, which means they have shapes that repeat, and musical instruments generate these periodic soundwaves for every note in their range. A single wavelength consists of a peak (compression of the air molecules) and a trough (rarefaction), and physicists measure wavelength in degrees (as though the wave had been compressed into a circle). A complete cycle lasts 360°, with the peak at 90° and the trough at 270°. Scientists define the time a wave takes to complete one cycle as its period. The simplest waveform, the sine wave, has the shape seen in Figure 1.2.

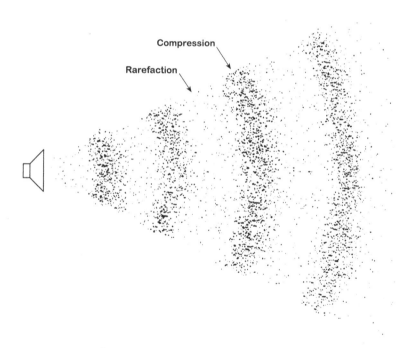

Figure 1.1 Compression and rarefaction of soundwaves.

Physicists usually represent soundwaves in this way, that is, by undulating lines on a graph, instead of by drawings that show the actual compression and rarefaction of air molecules (as seen in Figure 1.1). Figure 1.3 demonstrates the relationship between the physical phenomenon of sound and the graphic representation commonly used to depict soundwaves.

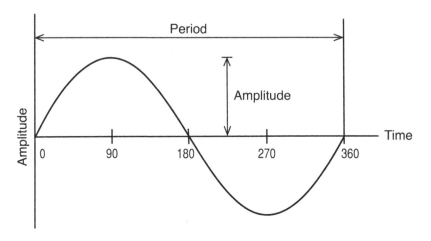

Figure 1.2 Periodic waveform.

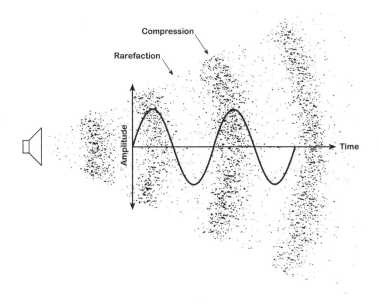

Figure 1.3 Graphic representation vs. the physical phenomenon of sound.

The number of complete cycles per second determines the frequency of a sound, and physicists state this number in hertz (Hz). A frequency of 1,000 Hz (1 kHz) means that the wave repeats 1,000 times every second, with each cycle lasting 1 millisecond (ms). Figure 1.4 shows two sine waves of the same amplitude (level) but with different frequencies. The wave occupying period T_1 (the solid line) is half the length of the wave occupying period T_2 (the dotted line). Hence, if T_1 has a cycle of 1 ms and T_2 a cycle of 2 ms, the shorter wave would have a frequency of 1 kHz and the longer wave a frequency of 500 Hz.

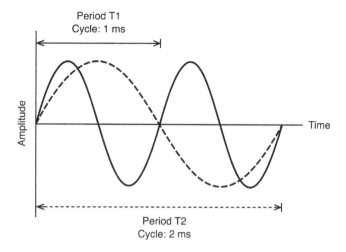

Period T1
Cycle: 1 ms

Amplitude

Time

Period T2
Cycle: 2 ms

Figure 1.4 Cycle and period.

The periodic nature of wave repetitions creates what we recognize as musical sound, but instruments would lack their characteristic timbral qualities if they generated only a single frequency for each note (that is, a single sine wave). Instead, they produce a complex set of frequencies arranged in a harmonic or overtone series above the lowest or fundamental frequency of the spectrum (the fundamental is the first harmonic of the series). We perceive the fundamental as the pitch of a note and the harmonics or overtones (also called partials) lying above it as the tonal color or timbre of that note. The sine waves shown above have pitch but they lack timbral quality, so physicists often describe them as pure tones. In other words, all complex waveforms with a recognizable harmonic timbre consist of a collection of sine waves, integer multiples of the fundamental frequency (two, three, four, etc. times the fundamental; see Figure 1.5).

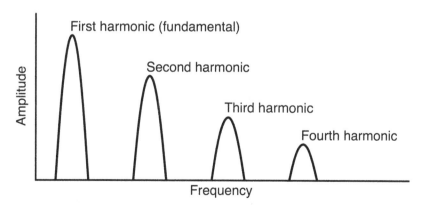

First harmonic (fundamental)

Second harmonic

Third harmonic

Fourth harmonic

Amplitude

Frequency

Figure 1.5 Harmonics. Note: complex waves without harmonic timbre (noise, for example) contain sine waves that are not integer multiples of the fundamental.

When expressed in terms of wavelength on a vibrating string, the mathematical relationship between the overtones and the fundamental can be depicted schematically, where a doubling of the frequency halves the wavelength, and so on (see Figure 1.6). The harmonic series can also be shown as notes on a staff (the blackened notes will be slightly out of tune in an equally tempered scale; see Figure 1.7).

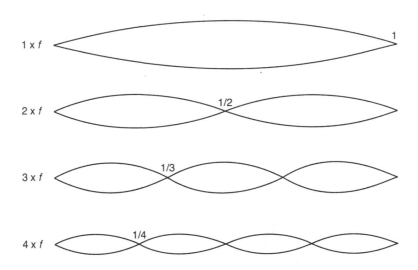

Figure 1.6 Mathematical relationship of overtones on a vibrating string (*f* = frequency).

Figure 1.7 Harmonic series shown in notes.

Instruments, then, receive their characteristic timbre from multiple frequencies sounding together, but because the amplitudes of the partials vary between instruments, timbre differs according to the relative strengths of specific overtones. Saxophones and clarinets, for example, emphasize the overtones shown in Figure 1.8, while Figure 1.9 shows the complex waveform for a single violin note, first as depicted in a DAW and then as a spectrogram revealing the overtone series lying above the fundamental.

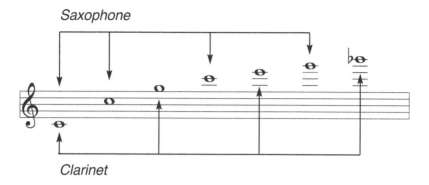

Figure 1.8 Timbre of clarinets and saxophones resulting from the relative strength of overtones.

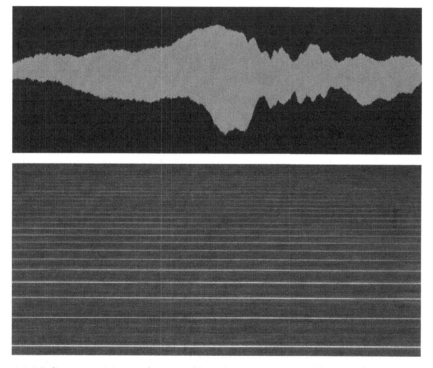

Figure 1.9 Violin note—(a) complex waveform; (b) spectrogram of fundamental and overtones.

Source: Screenshot from *Mixbus* used with the permission of Harrison Consoles (a). Screenshot from *Rx 6 Advanced* used with the permission of iZotope (b).

ENCLOSED SPACES

Reverberation

Soundwaves traveling in an enclosed space strike everything in the room, and unless the surfaces absorb the waves, the sound reflects. The term reverberation refers to the accumulation of random reflections arriving at a listening position so closely together that the hearer does not perceive each reflection separately.

Sound propagates spherically from a source in the direct or free field, decreasing in amplitude at a rate of 6.0 dB for every doubling of the distance (the Inverse Square Law; the drop is actually less in enclosed spaces, for the full decrease occurs only in purely free fields where sound propagation is undisturbed [outdoors] or in acoustically treated rooms that approximate free-field characteristics). A soundwave proceeding along the direct path, that is, the first wavefront, reaches the listener before the initial reflections from the floor, walls, and ceiling. The earliest identifiable reflections begin to arrive 30–80 ms after the first wavefront, and the size of the room, the nature of the surfaces, and the position of the listener determine the precise length of the delay (below 30 ms, hearers tend not to distinguish the reflections from the direct sound). As the soundwaves continue their travel, they repeatedly bounce off the room's surfaces and the reflections gradually increase in density. This denser wash of sound prevents listeners from hearing any of the reflections individually, and the transition from early reflections to reverberation starts to occur around 80 ms (the diagram in Figure 1.10 presents a simplified schematic of these phenomena).

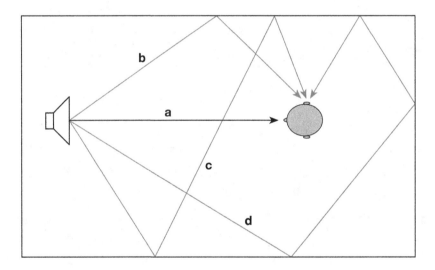

Figure 1.10 The reflection of soundwaves in a room (a = direct sound, b = first reflection, c & d = later reflections).

Furthermore, because the surfaces that the waves repeatedly strike absorb some of the energy of the late reflections, reverberant sound has a lower amplitude than that of the early reflections. Once a source has stopped emitting sound, the level of the late reflections gradually decays to a point where listeners no longer hear the reverberation. Acousticians define the time it takes for the sound pressure level of this complex set of room reflections to decrease to one-millionth of its original strength, a reduction of 60.0 dB, as the reverberation time (RT_{60}) of the space.

Figure 1.11 depicts the varying levels and density of complex room reflections. The dotted line represents the direct sound of the first wavefront, and in this example, the early reflections begin at 30 ms and continue until about 80 ms, after which the reflections gradually become reverberation; that is, they become dense enough that listeners cannot distinguish the individual reflections. Over time, this wash of sound fades to silence.

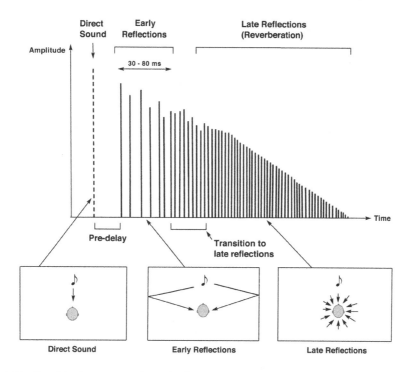

Figure 1.11 Graphical representation of reflections.

The reverberant quality of a room results from a mixture of direct sound and reflections, and these elements provide the information necessary for listeners to perceive depth and distance. In a large space, for example, greater distances between the hearer and the sound source cause the ratio of direct to reverberant sound to shift towards reverberation. In other words, as direct sound becomes weaker, reverberation becomes more prominent. The precise weighting of the two components helps listeners determine the location of a source and their distance from it. Direct sound provides hearers with information on the location of the source, whereas early reflections help them determine the size of the room. Late reflections, on the other hand, give the sound a sense of fullness.

Performance Venues

In studying concert halls and opera houses, acousticians have found that a number of factors contribute to what people regard as a "good" performance space, and after extensive interviews with conductors and music critics, Leo Beranek (2004: 2) compiled a set of desirable concert hall characteristics:

[A concert hall] must be so quiet that the very soft (*pp*) passages are clearly audible. It must have a reverberation time long enough to carry the crescendos to dramatic very loud (*ff*) climaxes. The music must be sufficiently clear that rapidly moving violin passages do not melt into a general "glob." The hall should have a spacious sound, making the music full and rich and apparently much "larger" than the instrument from which it emanates. It must endow the music with a pleasant "texture," an indescribable, but hearable, quantity that can be demonstrated by electronic measurements. The bass sounds of the orchestra must have "power" to provide a solid foundation to the music. Finally, there should be no echoes or "source shift"; that is to say, all or part of the orchestra should not seem to originate at some side or ceiling surface.

Of these attributes, reverberation figures prominently, as it fills the gaps between notes played in succession to give music "fullness of tone" (Beranek 2004: 21), and Beranek found that both performers and listeners prefer music from the Classical and Romantic periods to be heard in rectangular halls with reverberation times between 1.7 and 2.1 seconds (Beranek 2004: 2). This observation certainly matches the properties of some of the performance venues built in the past 250 years, but the exact type of space that might suit any given piece of music is probably determined more by the compositions themselves than by the predilections of modern audiences. For instance, organ music conceived for a European cathedral built in the late Middle Ages or Renaissance often needs the long reverberation times produced by those large spaces, while a Baroque concerto written for court performance might benefit from a smaller room with a much shorter RT_{60}.

Performance venues vary in their acoustic idiosyncrasies and may be divided into three main categories: those suitable for orchestral music, chamber music, and opera. Table 1.1 lists the size (in cubic meters), reverberation time (measured in seconds at mid-frequencies), and pre-delay (in milliseconds) for a number of venues from various countries. Orchestra halls tend to be large (generally more than 10,000 m^3), with reverberation times between 1.55 and 2.05 sec and pre-delays generally in the 20 to 25 ms range. Smaller chamber halls have correspondingly shorter reverberation times, averaging around 1.5 sec, with pre-delays below 20 ms. Opera houses, because of the need for intelligibility of text delivery, have the shortest reverberation times (around 1.2 sec) and pre-delays in the 15 to 31 ms range.

Table 1.1 Volume, reverberation time, and pre-delay of performance venues. The information in this table has been taken from Beranek 2004: 551, 619–25. Pre-delay times are given in Beranek's individual sections on the halls, as well as Table 4.4, p. 551. Beranek reports RT_{60} averages for mid-frequencies, 350–1400 Hz, of fully occupied halls.

	Volume (m³)	RT_{60} (sec)	Pre-delay (ms)
Orchestra Halls			
Amsterdam, Concertgebouw	18,780	2.00	21
Berlin, Kammermusiksaal der Philharmonie	11,000	1.82	20

	Volume (m³)	RT_{60} (sec)	Pre-delay (ms)
Boston, Symphony Hall	8,750	1.90	15
Brussels, Palais des Beaux-Arts	12,520	1.60	23
Leipzig, Gewandhaus	21,000	2.02	27
London, Barbican	17,750	1.68	25
New York, Carnegie	24,270	1.79	23
Philadelphia, Academy of Music	15,700	1.20	19
Salzburg, Festspielhaus	15,500	1.50	27
Vienna, Konzerthaus	16,600	1.88	23
Vienna, Großer Musikvereinssaal	15,000	2.00	12
Worcester (MA), Mechanics Hall	10,760	1.55	28
Zurich, Großer Tonhallesaal	11,400	2.05	14
Chamber Halls			
Amsterdam, Kleinersaal in Concertgebouw	2,190	1.25	17
Tokyo, Hamarikyu Asahi	5,800	1.67	15
Vienna, Brahmssaal	3,390	1.63	7
Vienna, Mozartsaal	3,920	1.49	11
Vienna, Schubertsaal	2,800	1.98	12
Zurich, Kleinersaal in Tonhalle	3,234	1.58	18
Opera Houses			
London, Royal Opera House	12,250	1.20	22
Milan, Teatro alla Scala	11,252	1.20	20
Naples, Teatro di San Carlo	13,700	1.15	31
Paris, Opéra Garnier	10,000	1.18	17
Sussex, Glyndebourne	7,790	1.25	20
Vienna, Staatsoper	10,665	1.30	15

But beyond these basic traits, other features of reverberant spaces influence the perception of sound quality. For example, early reflections can create a sense of intimacy, even in a larger hall, for when the pre-delay is short (about 20 ms) the sounds seem to originate from nearby surfaces, which gives a room the feeling of presence. This is one of the reasons Beranek cites (2004: 27–8) for the preference of rectangular-shaped performance spaces: because first reflections usually arrive from nearby sidewalls, narrower rooms have the lower pre-delays associated with intimacy. In wider spaces, however, first reflections occur later than in rectangular halls, so the sources tend to sound somewhat more diffuse.

Clarity also depends on the relationship between early and late reflections, for if the first reflections are louder than the reverberation, the "definition" or clarity of notes played in succession tends to increase. Moreover, narrow shoebox-shaped rooms contribute to an overall sense of spaciousness, for the abundance of early lateral reflections arriving at a listening position makes the instrument appear to emanate from a space considerably wider than that of the instrument itself (Beranek 2004: 29).

These factors clearly influence the perception of sound in reverberant spaces, so much so that some music seems better suited to certain types of rooms than others. As mentioned above, compositions with detailed contrapuntal lines (such as concertos and symphonies written in the eighteenth century) may benefit from narrower rectangular halls with strong early reflections and reverberation times of about 1.4 or 1.5 sec, whereas choral music conceived for the acoustics of large cathedrals may require a space that not only has proportionally more reverberant energy than early-reflection energy but also has an RT_{60} at or over 2.5 sec.

REFERENCE

Beranek, Leo. 2004. *Concert Halls and Opera Houses: Music, Acoustics, and Architecture*. 2nd ed. New York, NY: Springer.

CHAPTER 2

Audio Chain From Sound Source to Listener

Source—Mic—Amp—ADC—DAW—DAC—Amp—Speaker—Listener

INTEGRITY WITHIN AN AUDIO CHAIN

When listening to recordings, people hear analog copies of sound sources, and even though coloring of the sound can occur at every stage from microphone to loudspeaker, many recordists try to maintain a neutral audio path, so that the copy duplicates the original as closely as technology allows. However, since the audio signal passes through a number of devices as it travels along the chain, a certain amount of degradation is bound to occur when the small amounts of deterioration introduced by each device sum together.

In acoustic recording, a microphone makes an electrical copy of all the complex waveforms that strike its diaphragm, and this transduction turns those soundwaves into an analog electrical signal equivalent to what our eardrums would hear at the mic's location (that is, direct and reflected sound propagating from one or more sources). With the advent of computer-based recording in the early 1980s, a digital process now occupies a central position in the audio chain, and the way devices convert electrical current to and from digital information has become critical to preserving the integrity of the original signal.

BASIC CONCEPTS AND TERMINOLOGY

Overview of Signal Flow

The mechanical energy created by soundwaves striking a microphone's diaphragm induces an electrical current of constantly varying voltage that is analogous to the continuously changing air pressure of the waveforms that set the diaphragm in motion (positive voltage values indicate increase in pressure or compression and negative values, decrease in pressure or rarefaction). The weak current microphones produce, however, prevents the use of this signal in the next stages of the audio chain, so an amplifier within the mic generates line-level output that can be sent to an analog-to-digital converter (ADC) to change voltage into a digital form computers can recognize. This numeric information, the sound quality of which does not deteriorate when it is stored, copied, or processed in a computer, provides the basic material recordists manipulate in a digital audio workstation

(DAW). On leaving the DAW, the signal enters a digital-to-analog converter (DAC), so that the binary numbers can be decoded into variations of voltage. The resulting output passes through an amplifier before it reaches the loudspeakers. The speakers transduce the electrical current into mechanical vibrations that generate soundwaves which travel to the listener's eardrums.

Analog Audio

The term "analog" refers to the representation of a signal by continuously variable and measurable physical quantities, such as pressure or voltage. In acoustic audio recording, the measurement of constantly changing air pressure induces an electrical current, the voltage of which continuously varies up and down in the same way that the air pressure of soundwaves constantly increases and decreases (as a result, the fluctuation of an electrical signal's voltage corresponds directly to the amplitude variation of soundwaves). In a sense, electrons in wires perform the same function as air molecules. The conversion of soundwaves to electrical energy enables amplification, recording, editing, mixing, etc.

Digital Audio

Electrical current can be converted to digital information by using a series of discrete binary numbers (0, 1) to represent the changing voltage in an analog signal (fluctuating voltage equates directly to the amplitude variations of soundwaves). The conversion of analog voltage into digital information allows a signal to be much more accurately and reliably stored, processed, transmitted, and reproduced.

Pulse Code Modulation (PCM)

Invented by Alec Reeves in the late 1930s, PCM has developed into the standard method for digitally encoding analog waveforms (the technique is used in both WAV and AIFF). It has three components: sampling, quantizing, and encoding. During PCM, a device samples (measures) the voltage of an analog signal at a regular interval and then quantizes (rounds) those measurements to the nearest values on a pre-determined scale, each step of which has been assigned a discrete numerical value (whole number or integer). The device encodes these values in binary digits so that the original signal can be used in digital systems (see below for a fuller explanation of the main components).

Bit

The term bit abbreviates the expression "binary digit." Binary means something based on or made up of two things, and in digital audio systems, these two things are the numbers 0 and 1.

Bit Depth (Word Length)

Computers store data in number sequences that are multiples of eight digits, each of the numbers being either 0 or 1. A group of eight digits is called a byte, with one or more bytes comprising a binary word. 16-bit audio (the standard for CDs) means that every binary word contains 16 numbers; in 24-bit audio, each word has 24 numbers. Thus, bit depth stipulates how many 0s and 1s are used to represent each sample of a waveform.

In addition, every bit in a binary word equates to roughly 6.0 dB of dynamic range (dynamic range is the difference between the softest and loudest sound a system can produce), so in theory, the total dynamic range available on a CD is 96.0 dB (6.0 × 16), whereas 24-bit audio allows for a range of 144 dB, which exceeds the 130 dB humans can hear comfortably (today's technology, however, falls short of these theoretical limits by about 3.0 dB). In practice, though, the maximum dynamic range audio systems need to reproduce is the 120 dB or so orchestras can generate, and 24-bit audio easily surpasses what is required for these ensembles (in fact, a 21-bit system, with a theoretical dynamic range of 126 dB, would be adequate).

Bit Rate

The expression "bit rate" indicates how many bits are transmitted per unit of time in digital audio (the unit of time is often seconds, as in "bps" or "bits per second").

Analog to Digital Converter (ADC)

An ADC converts analog signals to digital code using pulse code modulation.

Sampling

Sampling is the process of measuring the voltage of an electrical audio signal at a regular interval so that the measurements can later be outputted as binary numbers. For example, when recording audio for CD distribution, waveforms are measured 44,100 times per second (44.1k), and the theory established by Harry Nyquist between 1924 and 1928 helps us understand why this sampling rate became the norm. Nyquist discovered that an analog signal can be recreated accurately only if measurements are taken at a rate twice the highest frequency in the signal (now known as the Nyquist frequency). In other words, the maximum frequency a digital system can represent is half the sampling rate.

Since 44.1k exceeds the minimum sampling rate of 40k required for the upper limit of human hearing (which is approximately 20 kHz), waveforms can easily be reconstructed accurately, and the excess capacity above 40k provides adequate space for reconstruction filters to operate. Sampling rates below that dictated by the Nyquist theorem prohibit the faithful restoration of signals, and a fault known as aliasing occurs when too few samples cause a device to interpret the voltage data as a waveform different from the one originally sampled. In Figure 2.1, dots represent the sampled points and the dashed line shows the incorrectly reconstructed waveform that results from too low a sample rate.

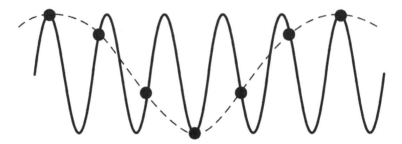

Figure 2.1 Aliasing.

Quantization

Sampling imposes a succession of discrete measurement points on a signal at a regular interval, and the non-continuous nature of these points means that digital systems cannot precisely copy the smoothly varying voltages of electrical current. A basic explanation of the principles converters use elucidates the process.

Because converters store each sample temporarily, holding it until the next sample is taken, the system ignores the continuously varying voltage between the sampling points. Beyond this loss of information, converters compare the measured voltage to a set of uniformly spaced values, akin to steps on a scale, instead of an infinite continuum of values, which means that a large range of precise measurements must be mapped onto a smaller set of whole numbers or integers (Christou 2008: 4–5). When the voltage at the sample falls between two values, the measurement has to be rounded or quantized to the closest number/step (see Figure 2.2). Hence, quantization changes the signal to match the points on the scale, a practice that introduces errors into the system (heard as nonrandom noise). The size of the error depends on the number of steps the scale contains: a 2-bit scale has four possible steps (2^2), a 3-bit scale eight steps (2^3), a 4-bit scale 16 steps (2^4), an 8-bit scale 256 steps (2^8), a 16-bit scale 65,536 steps (2^{16}), and a 24-bit scale 16,777,216 steps (2^{24}). Scales based on higher numbers of bits, then, because they have more finely graded values or steps, reduce the size of the rounding error and, hence, the amount of noise in the system. In fact, in both 16 and 24 bit scales, the rounding error is small, and this makes the noise created by quantization almost imperceptible (often 90 or more decibels below the signal) (iZotope 2011: 16).

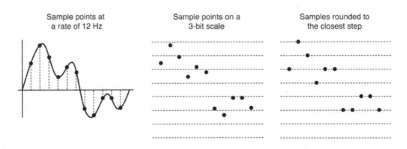

Figure 2.2 Rounding error in a 3-bit scale.

Nevertheless, many researchers believe it is "always desirable to reduce the audibility of any error in an audio signal" (Wannamaker 1991: 56), and to deal with modest amounts of distortion, engineers regularly employ dither when reducing audio from higher bit depths to lower ones. To decrease the number of bits from 24 to 16, a common procedure during the preparation of audio for storage or transmission, bit-depth converters discard eight of the least significant bits to shorten or truncate the binary word that represents the sampled voltage (that is, they remove the last eight digits of the binary word). This process, known as requantization (Wannamaker 2003: 2), rounds the samples to match a scale with fewer steps (65,536 instead of 16,777,216), which introduces the faint nonrandom noise mentioned above. But as the amplitude of the audio falls to its lowest levels, the relative size of the error becomes larger, and this means that during the quietest moments quantization/truncation can be audible. Researchers have found, however, that the addition of random noise to a signal replaces nonrandom distortion with a far more pleasing noise spectrum.

Indeed, psychoacoustic research has established that the more random the noise in a system is the less irritating it will be. Consequently, listeners find white noise (a type of random noise with a flat frequency spectrum) preferable to the nonrandom variety produced during quantization/truncation. The technique of dithering, then, adds specially constructed noise to the signal before its bit depth is reduced.

One of the commonly used types of dither is TPDF (triangular probability density function—white noise with a flat frequency spectrum), but devices or plugins can also add noise containing a greater amount of high-frequency content (called blue noise). The process involving blue noise, known as colored/shaped noise dithering or noise shaping, concentrates the distortion in less audible frequencies (generally those above 15–16 kHz), while reducing the level of the noise in the frequency range humans hear best (between 2.0 and 5.0 kHz and around 12 kHz) (iZotope 2011: 17–18). This form of dither sounds quieter than TPDF, and many engineers prefer it. Figure 2.3 presents a screenshot that shows

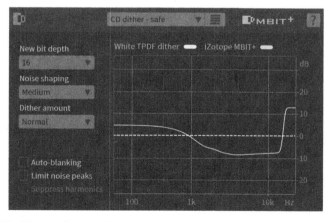

Figure 2.3 TPDF dither and noise shaping.

Source: Screenshot from MBIT+ used with the permission of iZotope.

both TPDF dither (the straight line at 0.0 dB) and noise shaping (the curved line that indicates the nature of the frequency attenuation and boosting).

But beyond this advantage, noise shaping increases the perceived dynamic range of 16-bit audio significantly, for the noise introduced by colored dithering can be as much as 118 dB below the peak level of the signal, and this has encouraged engineers to use some form of noise shaping whenever they lower the bit depth of audio from 24 to 16 (at and above 24 bits, the noise produced by quantization is insignificant, since it occurs at extremely low amplitudes).

One of the main benefits of 24-bit audio production is, then, to borrow the words of John Siau (Siau n.d.: unpaginated), Director of Engineering at Benchmark Media, "the ability to record and produce releases that can fully utilize the SNR [signal-to-noise ratio] available in a 16-bit system [16-bit audio has a SNR of 96.0 dB, and noise shaping can extend that dynamic range, at least perceptually, to 118.0 dB (iZotope 2011: 10–11)]." Christopher Montgomery of Xiph.org offers the following caveat:

> increasing the bit depth of the audio representation from 16 to 24 bits does not increase the perceptible resolution or "fineness" of the audio. It only increases the dynamic range, the range between the softest possible and the loudest possible sound, by lowering the noise floor. However, a 16-bit noise floor is already below what we can hear.
>
> (Montgomery n.d.: unpaginated)

Moreover, recording chains with naturally high levels of noise in them will receive no audible benefits from dithering, as the noise in the system will mask any improvements compensation for quantization error can offer.

Engineers apply dither as the last stage of preparing tracks for delivery, so that the added noise is decorrelated from the audio's signal (in other words, the replacement noise is not contained within the original signal). If recordists further process a track after dithering, the new plugins could, as Nugen Audio suggests, "over-accentuate the effect [of dither], resulting in audible noise artefacts" (Nugen Audio 2017: 10), especially if noise shaping had been part of the dither algorithm (EQ adjustments in the higher frequencies, for example, could make a recording sound too bright).

Digital to Analog Converter (DAC)

A DAC converts digital code to an analog signal (voltage), so that non-digital systems can use the information. The device changes the numeric encoding back to the voltage points the binary words represent, and this results in an electrical current comprised of single-point voltage levels. The DAC then sends this discrete waveform through a reconstruction filter to interpolate (fill in) the missing data to restore the signal to smoothly varying voltages.

Resolution

The sound quality of digital audio primarily depends on two elements in the system, sample rate and bit depth. Sample rates determine how frequently the waveform is measured and bit depth governs the number of binary digits used to store the data. Higher sample rates increase the frequency bandwidth devices can encode, while larger bit depths expand the dynamic range and reduce the noise in the system. For many engineers, these two components determine the "resolution" of the audio, and today "high-resolution audio" has a bit depth of at least 24 and a sample rate at or greater than 88.2 or 96 kHz.

REFERENCES

Christou, Cameron N. 2008. "Optimal Dither and Noise Shaping in Image Processing." M.Sc. thesis, University of Waterloo.

iZotope. 2011. *Dithering with Ozone: Tools, Tips, and Techniques*. Cambridge, MA.

Montgomery, Christopher. n.d. "24/192 Music Downloads . . . and Why They Make No Sense." Available at: https://people.xiph.org/~xiphmont/demo/index.html.

Nugen Audio. 2017. *ISL 2, Operation Manual*. Leeds, UK.

Siau, John. n.d. "Bit Depth." Available at: https://benchmarkmedia.com/blogs/application_notes/14949345-high-resolution-audio-bit-depth.

Wannamaker, Robert A. 1991. "Dither and Noise Shaping in Audio Applications." M.Sc. thesis, University of Waterloo.

Wannamaker, Robert A. 2003. "The Theory of Dithered Quantization." Ph.D. dissertation, University of Waterloo.

PART 2

Production

CHAPTER 3

Microphone Types

A microphone measures variations in air pressure and transduces soundwaves from acoustic energy to mechanical energy to electrical energy. Ideally, the signal a microphone produces should replicate the waveforms entering the mic as faithfully as possible, so that when a loudspeaker converts the signal back to acoustic energy, listeners hear a reasonably accurate representation of the sound source.

THE BEHAVIOR OF A PURE DIAPHRAGM

If a thin plate, or diaphragm, is suspended between two points in a way that exposes both the front and the back to soundwaves, it responds to any difference, or gradient, in air pressure between the two sides. For example, when soundwaves arrive at the front, they exert more pressure on the plate than exists behind it, and this difference causes the diaphragm to flex towards the rear (the opposite happens when soundwaves strike the back). But if soundwaves arrive at the edge, the pressure on the front and back remains identical, and this prevents the diaphragm from moving. Hence, pressure differences/gradients drive the diaphragm, and as shown in Figure 3.1, a freely suspended pressure-gradient transducer naturally produces a bi-directional or figure-8 pickup pattern; that is, it responds equally to sound arriving at the front or the back (0 and 180 degrees)

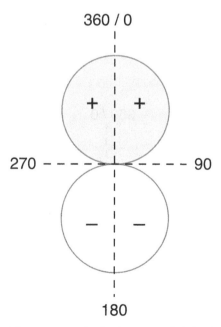

Figure 3.1 Bi-directional pickup pattern of a freely suspended diaphragm.

but does not react to sound approaching from the sides at 90 and 270 degrees (the "null" areas). When manufacturers enclose diaphragms, they place these thin plates in the sorts of specialized environments (capsules) that produce the pickup patterns commonly used today.

CONDENSER MICROPHONES

Condensers operate electrostatically. The capsule that turns mechanical vibration into electrical signal consists of a movable diaphragm and a fixed backplate, which form the two electrodes of a capacitor (previously called a condenser; hence, the name) that has been given a constant charge of DC voltage by an external power source (often supplied from a pre-amp and called phantom power). As sound-waves strike the diaphragm, the distance between the two surfaces changes, and this movement causes the charge-carrying ability (capacitance) of the structure to fluctuate around its fixed value. The resulting variation in voltage creates an electrical current that corresponds to the acoustic soundwave. A vacuum tube (valve) or transistor then boosts the current to ready the signal for post-microphone amplification through an external pre-amp.

Condensers employ either a pressure or a pressure-gradient principle of operation.

Pressure Transducer (Omnidirectional Polar Pattern)

In pressure transducers, microphone designers clamp a single circular diaphragm inside a completely enclosed casing so that they expose only the front face to the sound field. Soundwaves arriving from all directions exert equal force on the diaphragm, and because the diaphragm responds identically to every pressure fluctuation on its surface, these mics exhibit a non-directional, that is, an omni-directional (360°), response pattern (see Figure 3.2).

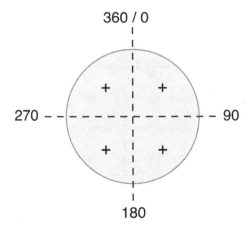

Figure 3.2 Omnidirectional pickup pattern.

Manufacturers often use polyethylene for the diaphragm and thinly coat one side with a metal, such as gold. Small holes, evenly distributed across the backplate, dampen the diaphragm's motion by capturing air as the diaphragm flexes in one direction or the other (see Figure 3.3). A narrow tube connects the interior chamber to the exterior so that the internal and external air pressure remain equal under any atmospheric condition.

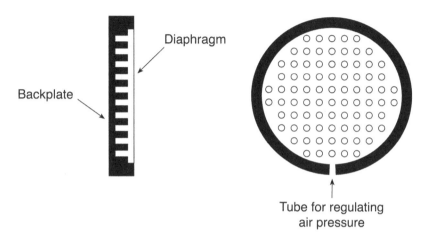

Figure 3.3 Diaphragm and backplate of a pressure transducer.

For use in the free or direct field, omnis can be designed to exhibit a fairly flat frequency response, as the mic mainly picks up direct sound (which has not lost high-frequency content). When recordists use omnis in the diffuse or reverberant sound field, however, the air in the room, as well as the various surfaces the soundwaves strike, absorb some of the higher frequencies. Manufacturers compensate for this loss by altering the design of free-field omnis to make the capsule more sensitive to higher frequencies, so that less of the high-frequency detail is lost.

Pressure-Gradient Transducer (Directional Polar Pattern— Figure-8/Bi-directional; Cardioid and Its Derivatives, Supercardioid and Hypercardioid)

As discussed above, a single diaphragm suspended between two points operates on a pressure-gradient principle and naturally exhibits a bi-directional or figure-8 pickup pattern. However, by combining this pickup pattern with that of an omnidirectional pressure transducer, microphone designers can create unidirectional patterns. Many years ago, researchers realized that because the output of an omnidirectional mic is always positive (gray in Figure 3.2), while bi-directional mics have both positive (gray in Figure 3.1) and negative (white in Figure 3.1) lobes, they could place these capsules in a single housing and produce a pattern that rejected sound from the rear. This pattern resembled the shape of a heart, so these mics became known as cardioids.

Within the configuration, each capsule picks up sound arriving from the front (0°) in phase, and when engineers combine the two outputs electrically, the signal doubles (see Figure 3.4). Sound approaching from the sides, however, sets only the diaphragm of the omni capsule in motion, for figure-8 designs do not respond to soundwaves arriving at right angles. Consequently, the omni portion of the mic produces the side signal on its own. At the rear of the mic, the two capsules generate signals of opposite signs (omni, positive; bi-directional, negative), which cancel each other and make any sound approaching from the rear almost inaudible (the extent of the inaudibility depends, of course, on the angle of incidence). When represented in a single diagram, one can easily see both the summed output of the positive signals at the front of the mic and the cancellation at the rear (see Figure 3.5).

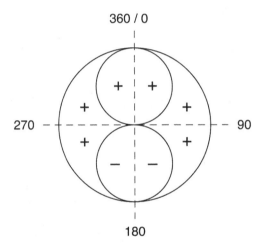

Figure 3.4 Bi-directional and omnidirectional patterns combined.

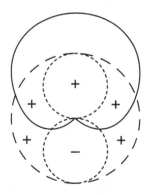

Figure 3.5 The three patterns superimposed (long dash = omnidirectional, short dash = bi-directional, and solid line = cardioid).

Although the first cardioid microphones contained two separate capsules within them, today's cardioids use principles of acoustic delay to achieve the same result from a single capsule. Manufacturers now place a delay network (also called a labyrinth) behind the diaphragm, and this altered path increases the time it takes for soundwaves to reach the back of the diaphragm; that is, external openings and internal passages in the housing force waves approaching from the rear to reach both the back and the front of the diaphragm at the same time, so that the waves cancel each other. Figure 3.6 shows a simple device, with a single side-entry slot, but designers achieve the desired delay in a number ways: multiple slots in the housing, plates with slots and holes, acoustic resistance (foam, for example), or a combination of methods.

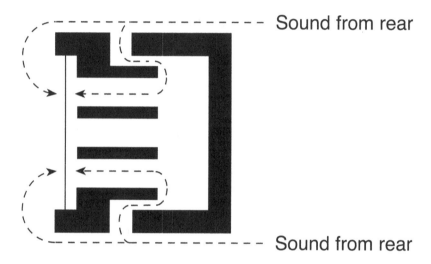

Figure 3.6 Acoustic delay in a cardioid capsule.

Thus, by introducing the correct amount of delay, capsules can attenuate soundwaves arriving from the rear. Moreover, manufacturers design unidirectional microphones to accept soundwaves from wider or narrower angles. Cardioids generally have a pickup or acceptance angle of about 131°, but this can be narrowed to 115° (supercardioid) and 105° (hypercardioid) (see Figure 3.7).

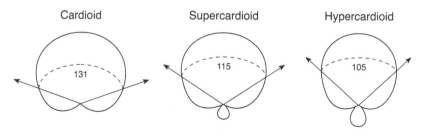

Figure 3.7 Acceptance angles of cardioid capsules.

Outside these acceptance angles, the ratio of direct to incidental sound changes enough for a noticeable loss of uniform response to occur (this change does not begin to become apparent until attenuation reaches about 3.0 dB, and microphone designers generally define an acceptance angle by the arc over which the sensitivity of a mic does not fall by more than 3.0 dB). Hence, a mic's response remains virtually identical within its angle of acceptance, and recordists should locate sources within the pickup angle if they wish to reproduce the sounds as a sonically homogeneous unit without noticeable degradation in tone color. As shown in Figure 3.7, cardioid microphones have the widest pickup arc among the three common unidirectional patterns. Bi-directional mics have a smaller acceptance angle of 90°.

Manufacturers also produce pressure-gradient electrostatic transducers with two diaphragms. In these double capacitors, each side of the device has a cardioid pattern, and by supplying voltages to the diaphragms independently, designers can achieve various response patterns in the same microphone (see Figure 3.8).

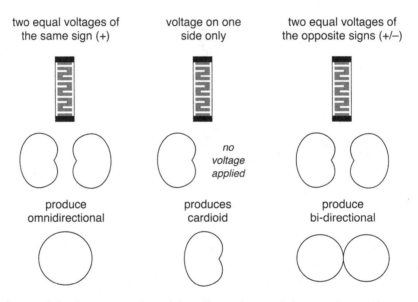

Figure 3.8 Dual diaphragm capsule and the patterns that result from applying voltage independently to each side.

DYNAMIC AND RIBBON MICROPHONES

Dynamic and ribbon mics operate on an electromagnetic principle in which some form of electrically conductive metal moves within a magnetic field to generate electrical current.

In dynamic mics, a light diaphragm connects to a finely wrapped coil of wire suspended in a magnetic field (see Figure 3.9). When soundwaves hit the face of the diaphragm, the attached coil begins to move within the magnetic field. This induces an electrical current proportional to the displacement velocity of the diaphragm (the greater the speed of motion, the greater the current); thus, dynamic mics are also called velocity or moving-coil microphones (these mics usually have a cardioid pattern).

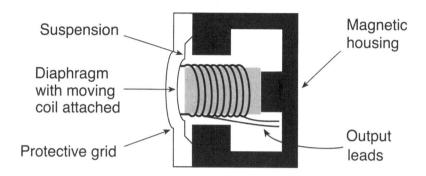

Figure 3.9 Dynamic microphone capsule.

Ribbon mics also operate on the principle of electromagnetic induction, but they produce their natural bi-directional pattern by suspending a diaphragm, consisting of an extremely thin strip of corrugated aluminum (the ribbon), in a strong magnetic field so that both sides engage with the sound source (see Figure 3.10).

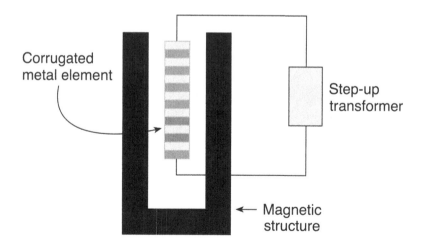

Figure 3.10 Ribbon microphone capsule.

Like dynamic mics, they induce an electrical current proportional to the velocity of displacement. The diaphragm, however, is narrow and short, with extremely low mass. This allows it to respond to the subtlest variations in sound pressure, but the lower conversion efficiencies characteristic of the design prevent the capsule from producing a strong current. Consequently, ribbon mics require considerable amplification before recordists can use the signal in an audio chain (engineers need pre-amps with a great deal of clean gain, as the self-noise of traditional ribbons tends to be well over 20.0 dB). However, a number of manufacturers now produce "active" ribbon mics that have much greater output than the older "passive" designs.

Microphone Characteristics

FREQUENCY RESPONSE

Ideally, microphones should respond equally well to all frequencies across the normal range of human hearing, 20–20,000 Hz. When expressed in a graph that plots output in decibels against frequency, this response appears flat. Although manufacturers find a flat frequency response difficult to attain in large spaces, they can come quite close to it in free-field applications, where direct sound predominates, as shown in Figure 4.1.

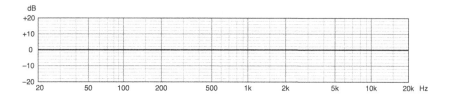

Figure 4.1 Flat frequency response.

However, in the diffuse field, where reverberant sound dominates, rooms tend to absorb treble frequencies, and manufacturers compensate for this phenomenon by enhancing a mic's response to higher frequencies. Microphone designers often boost the sensitivity of omnidirectional mics by 2.0–4.0 dB in the region of 10 kHz (the illustration in Figure 4.2 shows the typical response).

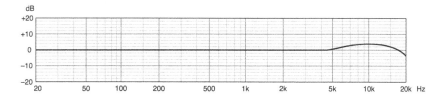

Figure 4.2 Sensitivity boost centered on 10 kHz.

DIRECTIONAL (POLAR) PATTERNS

The polar response of a microphone indicates its sensitivity to sounds arriving from any location around the mic. Manufacturers plot a capsule's response pattern on a polar coordinate graph comprised of 360° concentric circles, each circle representing a difference in sensitivity of a set amount (usually 2.5 or 5.0 dB). They locate the mic's diaphragm at the center and divide the circumference into 30° intervals, which allows them to show the relative sensitivity of the capsule for a given frequency in relation to the angle of incidence (mics respond differently at various frequencies, and the graphs show the amount of attenuation that occurs for specific frequencies in relation to 0° [on-axis]).

Omnidirectional microphones pick up sound equally from all directions, so they have a response pattern quite close to a perfect circle, at least for lower frequencies. The graphs in Figure 4.3 show the typical polar patterns of a small-diaphragm condenser. This mic has a purely omnidirectional pattern below 2.0 kHz, and it exhibits the normal narrowing of response at higher frequencies. This attenuation occurs because soundwaves approaching an omni mic from the rear with wavelengths equal to or less than the microphone's diameter tend not to bend around the end of the mic (see Figure 4.4; for a more detailed discussion, see Ballou 2008: 494–5). Consequently, at 4.0 kHz, the decreased sensitivity of the microphone amounts to just a decibel or so and is virtually unnoticeable, but between 8.0 and 12.5 kHz the drop certainly becomes apparent. Above 16 kHz, the response pattern of the mic approaches unidirectionality.

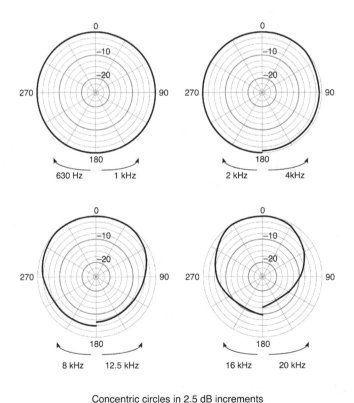

Concentric circles in 2.5 dB increments
Graphs divided into 30 degree intervals

Figure 4.3 Polar patterns at various frequencies for an omnidirectional microphone.

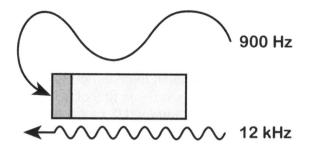

Figure 4.4 Attenuation of soundwaves approaching from the rear of a microphone, when the wavelengths are equal to or less than the microphone's diameter.

However, when recordists wish to restrict the amount of ambient or reverberant sound a mic captures, they generally employ directional models. These microphones can have both primary and secondary areas of pickup, as well as one or two null regions. Cardioid mics feature a reasonably wide pickup area at the front, with a large null at the rear (see the left diagram in Figure 4.5), while supercardioid and hypercardioid microphones restrict the width of their front lobes but have small lobes of limited sensitivity at the rear (the middle diagram in Figure 4.5). Although these latter two capsule types certainly reject sounds from behind, they do so mainly in the regions to the left and right of the rear lobe. Bi-directional mics have two lobes of equal size but opposite polarity, with nulls centering on 90° and 270° (the right diagram in Figure 4.5).

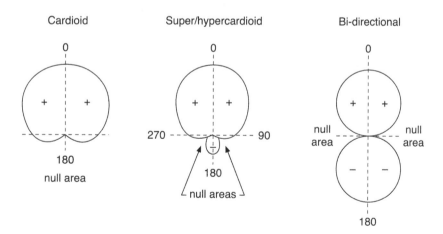

Figure 4.5 Pickup and null areas.

RANDOM ENERGY EFFICIENCY (REE; ALSO CALLED DIRECTIVITY FACTOR)

The efficiency of a mic can be determined by measuring the degree to which the capsule responds to the sound source itself (direct sound) relative to the reverberant field picked up from all directions. Omnidirectional mics respond equally to all sounds, and because of this, manufacturers use them as a reference against which they compare directional mics. Engineers assign omnis an REE of 1, and directional mics, which have a more "selective" response, an REE of less than 1. For example, both bi-directional mics and cardioids react to only about 1/3 of the total sound field, so they receive an REE of 0.333. This means that the ambient sound picked up by these mics is 1/3 or 4.8 dB lower than the direct sound. The narrower supercardioids and hypercardioids have REEs of 0.268 and 0.250 respectively. Supercardioids pick up 5.7 dB less ambient sound, whereas hypercardioids pick up 6.0 dB less.

DISTANCE FACTOR

Compared to an omnidirectional mic, recordists may place a directional transducer farther from a sound source and still produce similar audio results. Directional mics, because of their forward-oriented pickup patterns, reject a great deal of random off-axis sound, and when used in reverberant environments, engineers express the equivalent working distance of a directional microphone, relative to an omni, in terms of its distance factor, an indication of how far recordists can locate a directional mic from a sound source and have it pick up that instrument with the same ratio of direct to reverberant sound as an omni (see Figure 4.6). For example, the distance factor of a cardioid microphone is 1.7, which means that engineers may position it 1.7 times farther away from a sound source than an omni. This greater working distance results from the ability of directional mics to reject off-axis (reverberant) sound.

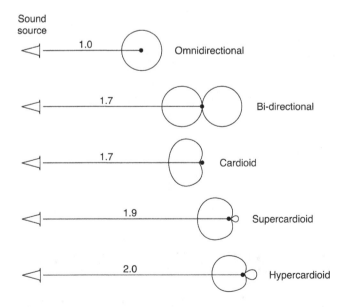

Figure 4.6 Distance factor.

In general, the distance between a microphone and its sound source can dramatically affect the character of a recording. When a recordist positions a mic close to a source, the microphone mainly captures direct sound, a placement that will exaggerate every sound emanating from the source, including imperfections. But if an engineer situates a mic quite a distance from the source, most of the direct sound will be eliminated. In this spot, the microphone "hears" the source as a whole but cannot capture subtle details. Between these locations lies the critical distance, a point at which the level of direct sound equals that of the reverberations (for a fuller discussion of room ambience, see Part 2, Chapter 6).

PROXIMITY EFFECT

All microphones that work on a pressure-gradient principle (bi-directional and cardioid patterns) exhibit proximity effect, a discernible increase in the low-frequency response as a sound source moves closer to the mic. Depending on the design of a microphone's capsule, the effect begins to become audible at distances of around 50 centimeters (20 inches) and is particularly noticeable under about 30 centimeters (12 inches).

A bit of physics, simplified for our purposes, helps to explain the effect. In the direct or free field of an open space (that is, a space free from reflections), soundwaves radiate in all directions from a source in ever-expanding spheres, and as the surface areas of these spheres increase over distance, the intensity of the sound decreases in relation to the area the soundwaves spread across (see Figure 4.7). Physicists define sound intensity as the amount of energy per unit of area (again, measured in terms of sound pressure level), and the Inverse Square Law states that the intensity of a sound decreases proportionally to the square of the distance from the source. In other words, when soundwaves radiate outwardly from a source in the direct field, they spread over a larger area, and for every doubling of the distance, the sound pressure reduces by half, which the human ear perceives as a decrease of 6.0 dB (somewhat less in enclosed spaces). The effect of the Inverse Square Law ends at the point in a room, called the critical distance, where direct sound and reverberation are equal in level (see Part 2, Chapter 6 for a fuller discussion of critical distance).

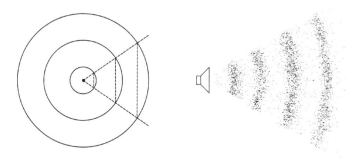

Figure 4.7 Spherical propagation of soundwaves in an open space.

When a recordist positions a microphone close to a sound source, the Inverse Square Law plays an important role in determining the relative pressures sound-waves exert on the front and back of the diaphragm, but when an engineer locates a mic farther from an instrument, the sound pressure on the front and back of the diaphragm approaches, for all practical purposes, equality, simply because the difference in distance that the soundwaves travel to reach either the front or the rear becomes negligible. The Inverse Square Law allows us to understand the principle: a 6.0 dB drop in loudness occurs only when the distance from the source doubles. For example, at 100 centimeters (1 meter), the path from the sound source to the diaphragm is significantly longer than the 1 centimeter path from the front of the diaphragm to the rear (the actual distance from the front to the back varies depending on the design of the capsule). Hence, an increase in the length of travel from 100 to 101 centimeters does not discernibly alter the sound pressure level on the rear side of the diaphragm.

But for shorter distances, the Inverse Square Law plays an important role in determining the relative pressures soundwaves exert on the front and back of the diaphragm, for the front-to-rear dimension is no longer insignificant. John Woram (1989: 102) has determined that a microphone with an internal path difference of 1 centimeter would have discrepancies in sound pressure level between the two sides of the diaphragm of 6.02 dB when placed 10 centimeters from the sound source and 0.83 dB at 100 centimeters. Figure 4.8 illustrates these principles in general terms.

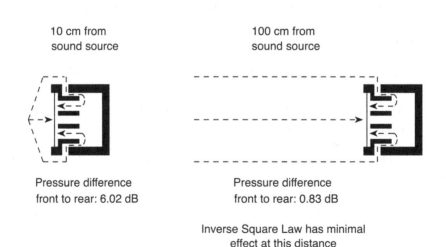

10 cm from
sound source

100 cm from
sound source

Pressure difference
front to rear: 6.02 dB

Pressure difference
front to rear: 0.83 dB

Inverse Square Law has minimal
effect at this distance

Figure 4.8 Inverse Square Law and discrepancies in sound pressure level.

Note that in Figure 4.8 a close mic placement resulted in 6.02 dB of greater sound pressure at the front of the diaphragm, whereas at 100 centimeters the difference between the front and the back shrank to less than 1.0 dB. Moreover, within the near field, the pressure differences between the front and the back of the diaphragm cause the sound pressure level to rise primarily for lower frequencies. This discrepancy results from phase anomalies across the frequency spectrum of a musical sound. The higher frequencies (shorter wavelengths) of a complex waveform do not bend around the edges of the capsule very easily and reach the two sides of the diaphragm proportionally more out of phase than lower frequencies (soundwaves across the normal range of human hearing vary in length from 1.7 centimeters [0.7 inches] at 20 kHz to 17 meters [56 feet] at 20 Hz). Consequently, a greater amount of cancellation occurs at higher frequencies. Low frequencies, however, consist of waves far longer than the diameter of the diaphragm, and because these waves bend around the capsule quite readily, they arrive at the two sides more in phase. Thus, the lower frequencies tend to sum together, and this summing boosts the low-frequency component of complex wave forms (the next section discusses phase more fully). The graph in Figure 4.9 shows how distance determines the degree of proximity effect below 640 Hz for a typical small-diaphragm condenser.

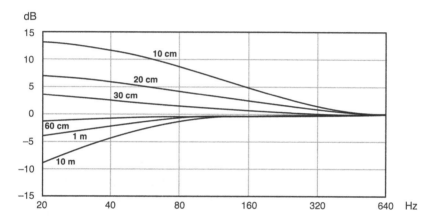

Figure 4.9 Proximity effect in a cardioid microphone.

PHASE

The term phase refers to the starting position of a periodic wave in relation to a complete cycle, and as mentioned above, physicists measure the cyclic nature of waves in degrees (see Figure 4.10).

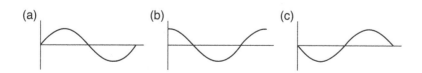

Figure 4.10 Phase in relation to a complete cycle: (a) starting at 0° (b) starting at 90° (c) starting at 180°.

If a soundwave (either simple or complex) arrives at a pair of microphones at the same time, that is, at an identical point in its cycle, the peaks and troughs align perfectly and combine electrically to produce a single waveform of double the amplitude or level (note that when an acoustic wave strikes a diaphragm, the transducer converts it into a similar wave shape within alternating current). Physicists refer to this doubling as constructive interference, and the two signals are "in phase," because they occupy the same relative position in time within a given cycle of the waveform (shown in Figure 4.11 as a single cycle of a sine wave).

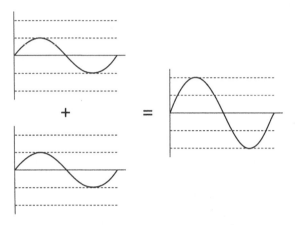

Figure 4.11 Constructive interference, "in phase."

But when a soundwave reaches the two microphones at different times, that is, at different places in its cycle, the peaks and troughs no longer align perfectly. Physicists call this misalignment destructive interference, and it can range from slight to complete. The combination of the two output signals retains the original frequency but at a lower level than an "in phase" configuration. A partial phase reinforcement/cancellation of a sine wave might look like the soundwave shown in Figure 4.12 (compare the lower amplitude in Figure 4.12 with the doubling shown in Figure 4.11).

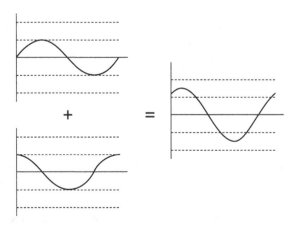

Figure 4.12 Destructive interference, partial reinforcement and cancellation.

Complete phase cancellation occurs when the peaks of one signal coincide exactly with the troughs of the other. Hence, a soundwave reaching one of the mics at the start of its cycle (0°) and the other at 180° produce a combined amplitude of zero (silence), as the two waves are, so to speak, 180° "out of phase" (see Figure 4.13).

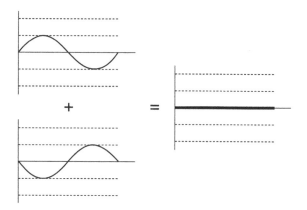

Figure 4.13 Destructive interference, complete cancellation.

The examples shown in Figures 4.10 to 4.13 represent just one frequency from a sound source, but because a spectrum of frequencies creates musical sounds (that is, complex waveforms consisting of multiple sine waves), constructive and destructive interference can occur simultaneously across a range of frequencies. For instance, if the delay in arrival at a second microphone matches the time it takes for one of the frequencies in a complex soundwave to complete a single cycle, a doubling of that frequency's amplitude results (a cycle of 1 ms combined with a delay of 1 ms doubles the amplitude of a 1 kHz frequency, for the wave arrives at the two mics "in phase"). But that same delay cancels the 500 Hz component of the spectrum, because the arrival times are 180° "out of phase." Other frequencies in the complex shift by amounts smaller than 180°, and these intermediate "phase shifts" partially cancel or reinforce the sine waves associated with those frequencies. In fact, short delays occurring in a complex waveform can produce comb filtering, a set of mathematically related (and regularly recurring) cancellations and reinforcements in which the summed wave that results from cutting and boosting frequencies resembles the teeth of a comb. The graph in Figure 4.14 illustrates the comb filtering produced when a signal reaches the second microphone of a pair 1 ms after it arrives at the first mic (peaks of 6.0 dB and dips of 30.0 dB are present).

"Three-to-One" Principle

These phase issues certainly have a considerable impact on the decisions engineers make about the placement of two or more microphones, for if recordists

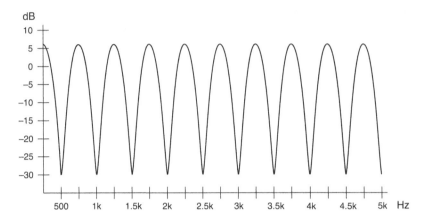

Figure 4.14 Comb filtering.

mistakenly capture out-of-phase signals, various degrees of interference, including comb filtering, can have quite a negative effect on sound quality, especially when phase-shifted signals are either played back or summed to mono. For example, the audibility of comb filtering (that is, the amount of undesirable "phasiness" or "hollowness" induced) depends on the time delay between the two mics: larger delays tend to render comb filtering inaudible, while shorter delays (particularly those under 10 ms) exacerbate the problem. Moreover, when the two mics have identical output levels, boosts of about 6.0 dB and cuts of as much as 30.0 dB frequently occur (as shown in Figure 4.14).

Engineers can, of course, minimize the negative effect of this distortion by reducing the difference in amplitude between the peaks and troughs caused by phase shifts. Phasiness becomes less prominent when the output level of one transducer is lower than the other, and an attenuation of 9.0 to 10.0 dB at one of the mics can decrease amplitude differences to about 4.0 dB, a level at which the worst offender, comb filtering, becomes tolerable for most listeners (some say it resembles a pleasant room ambience). In fact, by placing two microphones in accordance with the "three-to-one" principle, recordists can easily achieve 9.0 dB of attenuation.

Figure 4.15 demonstrates how audio researchers arrived at this principle through tests carried out in an anechoic chamber (the information in the following paragraphs has been taken from Burroughs 1974: 115–19). The technicians placed a sound source directly on-axis 2 feet (61 centimeters) in front of a microphone (number 1 in Figure 4.15) and recorded the sound through that mic. They then recorded the source five more times with two microphones, the second one located 2 feet (61 centimeters), 4 feet (122 centimeters), 6 feet (183 centimeters), 8 feet (244 centimeters), and 10 feet (305 centimeters) away from the on-axis mic.

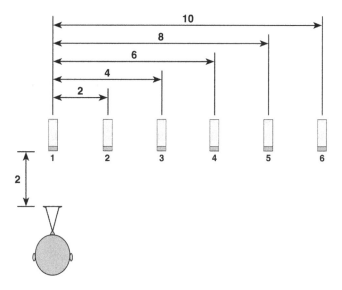

Figure 4.15 Testing procedure for the "three-to-one" principle.
Source: Adapted from Burroughs 1974: 117.

When plotted on a graph (see Figure 4.16), the on-axis response of microphone number 1 in Figure 4.15 produced the curve shown at distance "0" (the top contour). The blended signals from the two mics generated the frequency-response curves given below it (the numbers at the ends of the contours indicate the distance between the microphones).

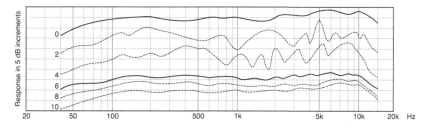

Figure 4.16 Frequency response curves.
Source: Adapted from Burroughs 1974: 118.

These curves illustrate the varying amounts of deterioration that combined signals can cause. When the researchers located the second transducer either 2 feet (61 centimeters) or 4 feet (122 centimeters) from the on-axis mic, noticeable reinforcements and cancellations occurred, and only when they placed the second microphone 10 feet (305 centimeters) away did the combined signals parallel the on-axis response of mic number one. However, the response curves at 6 feet (183 centimeters), 8 feet (244 centimeters), and 10 feet (305 centimeters) resemble one another, and despite the subtle differences in detail that the scientists measured in the anechoic chamber, skilled listeners did not notice audible improvements in sound quality when the researchers positioned the second microphone farther than 6 feet (183 centimeters) away. At 6 feet (183 centimeters), the distance between the two microphones is three times as great as the 2-foot (61-centimeter) distance from the on-axis mic to the sound source. Thus, the test revealed that the engineers did not need to have more than a three-to-one ratio between mics to avoid noticeable phase interference.

In a multiple microphone setup, recordists can establish this ratio with the distances (all in feet) indicated in the diagram shown in Figure 4.17 (in practice, one may need to modify the placement somewhat, for the theoretical model does not take into account differences in loudness between the sound sources).

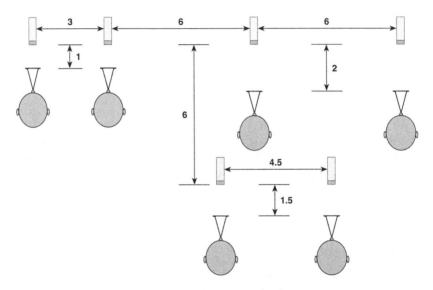

Figure 4.17 Multiple microphone setup to avoid comb filtering.
Source: Adapted from Burroughs 1974: 118.

These locations should result in a difference of 9.0 to 10.0 dB between microphones; that is, the background sounds picked up by any given mic should be at least 9.0 dB lower than the primary sound source in front of the microphone. During rehearsal, recordists can check this attenuation one mic at a time. With every microphone switched off except for one, the level at the sole remaining mic should drop by at least 9.0 dB when the performer at that microphone stops playing (in other words, the mic should "hear" the background sounds at a level 9.0 dB lower than the sound produced by the player directly in front of the microphone).

In summary, if recordists wish to maintain sufficient phase integrity in concert/recital hall locations, they simply need to follow the "three-to-one" principle: for every unit of distance between a microphone and its source, separate nearby mics by at least three times that distance. However, in situations where engineers opt for close miking and/or further processing that requires signal splitting, recordists regularly correct small amounts of phase shift during post-production either by manually moving the signals to synchronize them or by employing a plugin, such as Sound Radix's *Auto Align*, to bring the signals into an acceptable phase relationship.

Auto Align analyzes pairs of signals to find the amount of time delay between them and then compensates for the difference (see the screenshots in Figure 4.18). The plugin presents its measurement in samples, milliseconds, and distance (centimeters and inches), and beyond correcting out-of-phase signals, it allows engineers to "enhance the sense of space" when some form of delay is desirable. In other words, *Auto Align* can "time-place the [distant] microphones to better match the close-mic'd source" (Sound Radix n.d.: 2).

Figure 4.18 *Auto Align*, two signals compared and phase-aligned (72 sample delay between them).

Source: Used with the permission of Sound Radix.

REFERENCES

Ballou, Glen M., ed. 2008. *Handbook for Sound Engineers*. 4th ed. Boston, MA: Focal Press.

Burroughs, Lou. 1974. *Microphones: Design and Application*. Plainview, NY: Sagamore Publishing.

Sound Radix. n.d. *Auto Align, User Manual*. Tel Aviv.

Woram, John M. 1989. *Sound Recording Handbook*. Indianapolis, IN: Howard W. Sams.

CHAPTER 5

Stereo Microphone Techniques

Phase integrity obviously bears directly on recording in stereo, and over the decades audio engineers have developed a number of successful strategies for producing realistic stereo images, some of which use principles of time lag to achieve the desired effect.

Stereo miking allows engineers to replicate the way listeners hear individual instruments or ensembles in concert and recital halls, and recordists often employ just two or three microphones to capture the soundwaves emitted by the instrument/ensemble. When reproduced on a stereo playback system, the transduced sound energy generates phantom images at various points between the two loudspeakers. These images simulate the locations of the instrument(s) on the original sound stage and create the illusion of space.

Listeners use two cues, intensity (measured in terms of sound pressure level) and time of arrival, to locate the origin of sounds on a horizontal plane in front of them. Sounds which either arrive first or are stronger provide directional information for listeners, and recordists rely on time and level differences to simulate the sense of spaciousness audiences experience at a "live" event.

In stereo playback systems, listeners hear just two sources of sound, one left and one right, and the phantom images people perceive between the loudspeakers create the illusion of a stereo panorama. The optimum angle for stereo perception is 30° from an imaginary center line to each loudspeaker, and this arrangement places both the listener and the speakers at the corners of an equilateral triangle (see Figure 5.1). When the playback system feeds an identical signal to the loudspeakers, the same amount of sound energy radiates from each one, and the listener perceives the sound not as coming from the individual speakers but from a point equidistant between them (the phantom image). Thus, in this physical layout, no intensity (sound pressure level) or time-of-arrival differences exist at the listening position.

Fortunately, psychoacoustic researchers have established the range of level and time discrepancies required for listeners to perceive sounds as coming from various points along the horizontal plane between left and right loudspeakers. If no level or time differences exist, the hearer localizes the sound source at 0° (as noted above and shown in Figure 5.1), but if recordists want the listener to locate the source somewhere between 0° and 30°, they would have to introduce variations in level or time (or both) between the left and right channels. A difference of 15.0 to 20.0 dB or a delay of about 1.5 ms causes the phantom image to shift all the way to one of the speakers, while smaller variations localize images at specific points along the horizontal plane (see Figure 5.2).

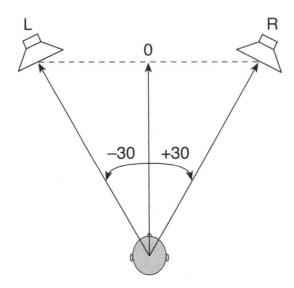

Figure 5.1 Optimum position for listening to stereo playback.

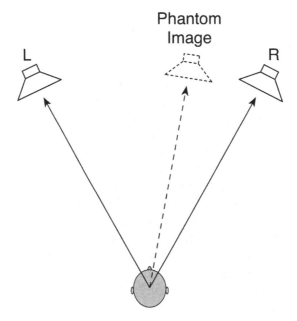

Figure 5.2 Phantom image in stereo playback.

Stereo miking uses the same principles of level and time-of-arrival differences to generate phantom images that simulate what a listener, seated in an ideal location, hears at a "live" event. But, of course, if an engineer wishes to localize instruments appropriately along the plane from left to right, the sound sources must fall within the acceptance angle of the pair of microphones, for if the outer instruments lie beyond the pickup angle, the sound sources will not be localized properly on stereo playback.

COINCIDENT PAIRS

By placing two microphones as closely together as possible, audio engineers can virtually eliminate time lags between the mics so that they achieve stereo imaging exclusively through level differences. Such configurations, called coincident pairs, align the diaphragms of the microphones vertically above one another, and this enables the mics to "hear" sound sources along a horizontal plane as if their diaphragms were in the same place. Three manifestations of this concept have become popular among recordists, X-Y, Blumlein, and mid-side.

X-Y

In an X-Y configuration, recordists generally use a matched pair of directional microphones (normally cardioid) and position them symmetrically around an imaginary center line, one mic angled to the left side of the ensemble and the other to the right (see Figure 5.3).

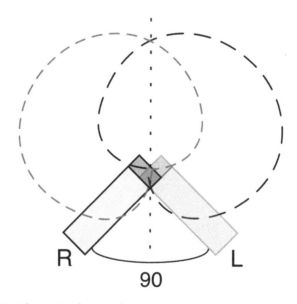

Figure 5.3 X-Y coincident pair of microphones.

Because directional mics exhibit the greatest sensitivity to sounds arriving on-axis and become progressively less sensitive to off-axis sounds, they produce a significantly higher level of signal from sound sources directly in front of them and lower level signals from sources in other locations. When engineers place two cardioids in the way shown in Figure 5.3, each microphone generates the same amount of signal for instruments in the center of the ensemble. On playback, the identical energy radiating from the two loudspeakers produces a phantom image midway between the speakers. But instruments situated some distance from the center will be more on-axis for one mic than the other. When, for example, a performer stands closer to the right-facing microphone, the right mic generates more signal than the left for that instrument. On playback, a higher level radiates from the right loudspeaker of the stereo system than the left, so that the instrument appears off-center to the right, the location of the musician in the original sound stage.

The X-Y configuration produces a strong sense of lateral spread across the stereo sound stage, with a high degree of phase integrity. Although an angle of 90° between the mics remains common, recordists actually place them anywhere from 80° to 130°. The width of the ensemble generally determines the placement, for larger angles between the axes of the microphones create not only wider recording angles but also broader stereo images during playback. Typically, recordists place coincident pairs back from the ensemble about half the width of the sound stage to capture the performers from more of a reverberant perspective.

Audio engineers also use omnidirectional mics in X-Y configurations, and recordists find that a 60° to 90° angle produces a clear central image with a sense of stereo space. Omni pairs have the advantage of maintaining a relatively stable image even when a soloist moves from side to side.

Blumlein

In the 1930s, the British inventor, Alan Blumlein, developed the coincident technique which bears his name. For this array, Blumlein used two bi-directional microphones set at an angle of 90°, so that the axis of maximum sensitivity in one of the mics aligned directly with the axis of least sensitivity in the other (see Figure 5.4). This makes each microphone relatively insensitive to sounds coming from the opposite direction, and the arrangement produces considerable separation between the signals.

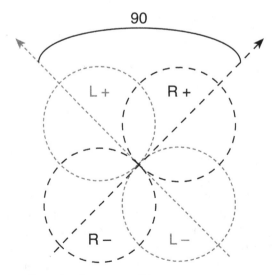

Figure 5.4 Blumlein configuration for stereo miking.

On playback, sound arriving at the 0° axis of the pair creates a strong phantom image midway between the loudspeakers (primarily because sounds located centrally between the pair of microphones are actually off-axis by the same amount for each mic). But when engineers place sound sources to the right or left of the 0° axis, the output of one mic increases and the other decreases. At a 45° angle of incidence, either the right or the left microphone reaches its maximum output, while the other remains at zero (for the sound arrives at the null of this mic). Consequently, the varying degrees of level differences between the microphones precisely track the angles of sound arrival. During playback, listeners hear what some people describe as the most accurate representation of the original sound stage.

In many recording situations, the rear lobes of a Blumlein pair face the ambient space of the hall, and when engineers position the mics within the critical distance, they can achieve an appropriate balance between direct and reverberant sound. However, in a hall with poor acoustics, the pair also captures the unfavorable nature of the room. One other disadvantage of the Blumlein technique concerns side reflections. Since both the positive and the negative lobes of the mics pick up lateral reflections, phase cancellations between the two may result in a "hollow" sound quality for listeners.

Mid-Side (M/S)

This technique, developed in the 1950s by Holger Lauridsen of Danish State Radio, uses two different polar patterns to produce a stereo image. A cardioid mic pointing directly at the center of an ensemble picks up the middle of the group (M), while a bi-directional microphone, pointing laterally with its null at the 0° axis of the cardioid, captures the sides (S) of the ensemble (see Figure 5.5).

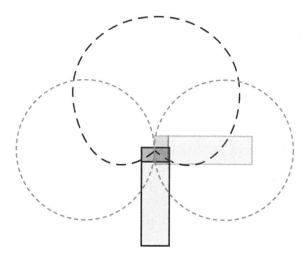

Figure 5.5 Mid-side technique.

Engineers must process the signals generated by the microphones to produce left and right channels suitable for normal stereo listening. When summed, the superimposed images of the array shown in Figure 5.5 create two polar patterns, one oriented to the left and the other to the right (see Figure 5.6).

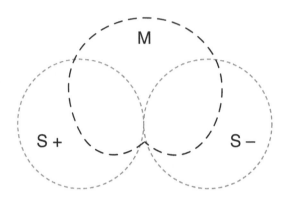

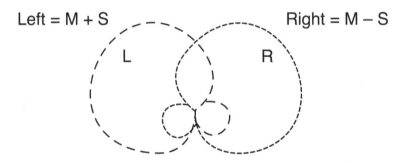

Figure 5.6 Summing of the mid-side pickup patterns.

The M/S technique allows engineers to manipulate the two signals by electrical/digital means, and this gives them a great deal of flexibility in creating a stereo image. By varying the amount of the mid and side components, they can easily change the width of the stereo sound stage. For example, a larger amount of mid signal narrows the sound stage, whereas more side signal broadens it.

Engineers achieve this variability by using three input channels of a mixer. They send the signal from the mid mic to its own channel, panned center, and the signal from the side microphone to two separate channels, panned left and right respectively, with the phase/polarity of the right channel inverted. By adjusting the controls in the mixer, they can alter the width of the stereo image (see Figure 5.7).

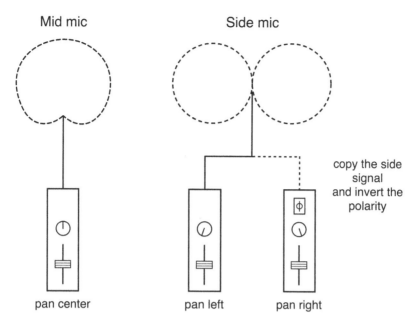

Figure 5.7 Mid-side signal flow in a mixer.

NEAR-COINCIDENT ARRAYS

Over the years, audio engineers have developed several strategies for using microphones to mimic human hearing. When recordists separate the capsules of two mics by distances that approximate the spacing between the ears of the average person, they can achieve stereo images through differences in both intensity and time of arrival. This near-coincident technique adds a sense of spaciousness to recordings, for it introduces a small degree of phase interference between the microphones in the higher frequencies (that is, comb filtering above about 1 kHz). The arrays commonly employed today are known by the acronyms ORTF, NOS, DIN, and OSS, and they tend to produce a wider stereo image than the classic X-Y arrangement.

ORTF

Named after the Office de la Radiodiffusion-Télévision Française (French Radio-Television Office), where the technique was developed, ORTF produces a pleasing stereo image, while maintaining adequate phase integrity for monophonic broadcasting. In this array, recordists angle two cardioid microphones at 110°, with the capsules separated by 17 centimeters (6.7 inches). The configuration approximates the average distance from one ear to the other, as well as the typical reception angle (see Figure 5.8).

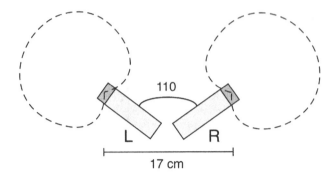

Figure 5.8 ORTF configuration of stereo microphones.

The use of cardioids provides significant differences in level between the two channels, and at low frequencies, the signals from the mics are virtually phase coherent. At higher frequencies (1 kHz and above), the slight amounts of comb filtering present in the combined signals create what many describe as a sense of "air" or "openness" in the stereo image. A number of recordists feel that this array produces, in the words of Bruce and Jenny Bartlett, "the best overall compromise of localization accuracy, image sharpness, an even balance across the stage, and ambient warmth" (Bartlett 2007: 217–18).

NOS

Audio technicians at the Nederlandse Omroep Stichting (Netherlands Broadcasting Foundation) developed another type of near-coincident array. In this configuration, engineers place two cardioid mics at an angle of 90° and separate the capsules by 30 centimeters (11.8 inches) (see Figure 5.9). This technique relies primarily on differences in level between the microphones, but the wider spacing does produce phase incoherence that begins to become audible at about 250 Hz. This makes the technique less useful for monophonic broadcasting.

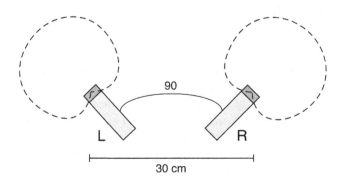

Figure 5.9 NOS configuration of stereo microphones.

DIN

Engineers at the Deutsches Institut für Normung (German Institute for Standard-ization) experimented with another variation of the near-coincident array. They placed two cardioid microphones at an angle of 90° and separated the capsules by 20 centimeters (8 inches) (see Figure 5.10). This technique produces a stereo image through a blend of level and time-of-arrival differences, and recordists find it particularly useful at shorter distances, especially for pianos, small ensem-bles, or sections of an orchestra.

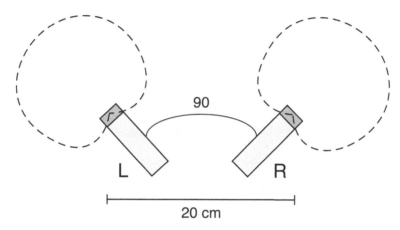

Figure 5.10 DIN configuration of stereo microphones.

OSS (Optimum Stereo Signal; Also Known as the Jecklin Disk)

This system, first proposed by the Swiss radio engineer Jürg Jecklin in 1981, approximates natural binaural (human) hearing by simulating the differences in level, time, and frequency response listeners experience at a "live" event, but in a way that reproduces a realistic stereo image through loudspeakers (note that the binaural approach to stereo recording, a technique which places omnidirec-tional mic capsules in the ears of a dummy head, only produces a clear stereo image when the two channels are reproduced through headphones).

Jecklin's original technique (Jecklin 1981) places an omnidirectional micro-phone on either side of a round disk (28 centimeters or 11 inches in diameter) covered with absorbent material to reduce reflections from the disk (see Fig-ure 5.11). He positioned the capsules 16.5 centimeters (6.5 inches) apart and angled them out from the center of the disk by 20° to replicate the ear positions and soundwave incidence of the average human head. Jecklin later expanded the dimensions of his technique, recommending a 36 centimeter (14.25 inch) spac-ing between omnidirectional mics mounted on either side of a 35 centimeter (13.75 inch) disk.

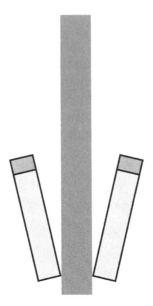

Figure 5.11 Jecklin's OSS technique.

The acoustic baffle between the mics improves the apparent width and clarity of the stereo image. Below 200 Hz, both microphones receive the same signal, and as the frequency rises, diffraction at the edge of the disk increases the effect of separation (around 5.0 dB at 1 kHz and about 10.0 dB at 5 kHz).

The array produces a well-defined stereo image and works best with ensembles that can achieve internal balance (such as occurs in classical music) and in rooms that have nominal or short reverberation times. For the best ratio of direct to reflected sound, recordists often place the array at or within the critical distance in a location that produces an optimal sound quality.

SPACED MICROPHONES

In these arrays, audio engineers do not try to replicate the human head, and they place the microphones much farther apart than with near-coincident pairs. They also situate them well above the sound stage to reduce the differences in distance between the mics and the instruments situated at the front and rear of the ensemble. Recordists tend to favor omnidirectional microphones for these applications, primarily because omnis pick up sound equally well from all angles, which makes them ideal for achieving the desired proportion of direct to reverberant sound. The amount of reverberation recordists wish to capture determines where they locate the mics, and many engineers use the critical distance as a starting point for balancing room reflections and direct sound (the closer the placement the more the direct sound dominates; hence, the microphones pick up less reverberation).

A-B

In this configuration, engineers center two spaced microphones in front of an ensemble to create a stereo image through differences in arrival time. The requirements of the project determine the width of the spacing, but in general, recordists place the microphones an equal distance from the center line of the ensemble, often a third to half the distance from the middle to the outer edge (the greater the spacing, the wider the stereo image; see Figure 5.12).

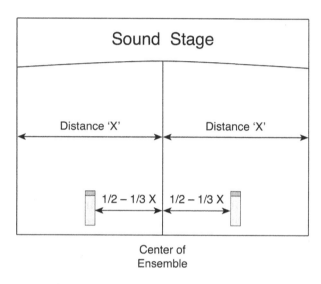

Figure 5.12 A-B stereo miking.

Sound sources located midway between the mics produce an identical signal in each transducer (because no level or time-of-arrival differences exist), and on playback, listeners hear the phantom images of those sources halfway between the speakers. Sound sources situated to the left or right of this center line are closer to one of the mics than the other, and this prevents the soundwaves from reaching the microphones at the same time. When replicated in playback, the delays shift the phantom images off-center to simulate the location of the sources on the original sound stage. In fact, depending on the width of the sound stage, a microphone spacing of about 61 centimeters (2 feet) allows sounds at the far right and left sides of an ensemble to be reproduced accurately in those locations, because delays of about 1.5 ms, which this spacing creates, cause the images to shift all the way to one speaker or the other. Sound sources partly off-center produce inter-channel delays of less than 1.5 ms, and this allows listeners to perceive phantom images of those sources at various points along the horizontal plane between the loudspeakers.

For large ensembles (such as orchestras), engineers prefer a spacing of 3–4 meters (10–12 feet), but this can provide insufficient coverage for the center of the sound stage, leaving a "hole in the middle," so to speak. However, a third microphone, or another pair of omnis spaced about 61 centimeters (2 feet) apart, placed midway between the main pair, with the extra signal mixed into both channels, solves this problem.

While widely spaced pairs tend to produce clear center images, the human ear has more difficulty distinguishing off-center stereo images generated solely by differences in arrival time. Thus, in this technique, side images tend to sound unfocused (that is, listeners find them hard to localize), especially since random phase relationships between the two channels can cause comb filtering. This gives the overall sound of a recording a diffuse and blended character, instead of a sharp and focused one, but for many people, these defects create a sense of spaciousness in which concert hall reverberation seems to surround the instruments.

Faulkner Phased Arrays

In 1981, the British recording engineer Tony Faulkner wrote about his technique of spacing two bi-directional microphones 20 centimeters (8 inches) apart directly facing the sound source (Faulkner 1981; see Figure 5.13). The method produces "coherent imaging plus an open feeling" (Borwick 1990: 132), and Faulkner typically placed the mics a distance from the ensemble equal to or greater than the width of the sound stage. Recordists consider this location particularly useful for very reverberant environments, such as large churches, because the microphone pair allows them to obtain clarity at the center of the stereo image, what Faulkner refers to as a dryer sound (Røde 2011: 30:46–31:44), while capturing the natural ambience of the room. The increased distance from the ensemble also permits engineers to set the mics at a lower height than closer locations would allow (perhaps as low as ear level). But when recordists need to place the microphones nearer the ensemble, they sometimes use omnidirectional flanking mics (roughly a meter or 3.3 feet apart) to counter the narrow stereo spread that might result from the use of the spaced pair on its own.

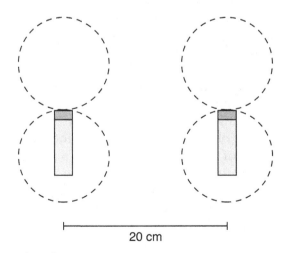

20 cm

Figure 5.13 Faulkner phased array.

Faulkner then expanded his phased array technique for use with large orchestras. Over the conductor's head, he places two omnidirectional mics 67 centimeters (26.5 inches) apart, with an ORTF pair of cardioids in the middle, 41 centimeters apart (see Figure 5.14). In this array, the near-coincident pair gives a sense of presence (closeness), while the omnis help create a feeling of ambience around the performers (Røde 2011: 38:55–39:04; Simmons 2016: unpaginated).

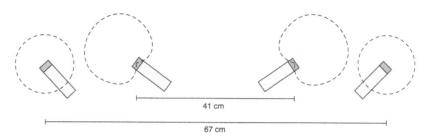

Figure 5.14 Expanded Faulkner phased array.

Decca Tree

This technique, originally pioneered by Roy Wallace of Decca Records in 1954, employs three omnidirectional microphones arranged in a triangle. The exact dimensions of the "tree" vary a great deal, but recordists often space the left and right microphones approximately 2 meters (6.5 feet) apart, with the center mic placed about 1.0 to 1.5 meters (39 to 59 inches) in front. In addition, they generally situate the array at or slightly behind and 2 or more meters (6.5 feet) above the conductor's head (see Figure 5.15).

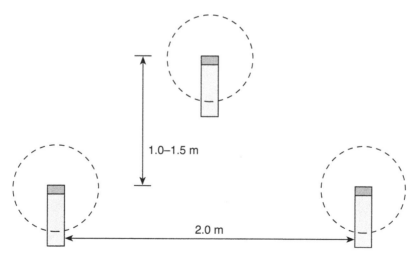

Figure 5.15 Decca tree.

The relatively close spacing of the left and right mics generates not only sufficient level differences to achieve good stereo imaging on playback but also enough phase information to create an open sound. By capturing a solid central image, the middle microphone eliminates the "hole-in-the-middle" problem. In short, the Decca tree produces a spacious sound with sharp images. Over the years since Roy Wallace first introduced this array, many variations of the technique have been employed. Recordists have altered the distances between the microphones, used cardioid and bi-directional mics, and replaced the center microphone with a coincident pair (often X-Y or M/S), a substitution that provides "articulation" for the stereo image, while generating a sense of "spaciousness" through the outer mics.

REFERENCES

Bartlett, Bruce and Jenny Bartlett. 2007. *Recording Music on Location: Capturing the Live Performance.* Boston, MA: Focal Press.

Borwick, John. 1990. *Microphones: Technology and Technique.* Boston, MA: Focal Press.

Faulkner, Tony. 1981. "A Phased Array." *Hi-Fi News & Record Review* 26: 44–6.

Jecklin, Jürg. 1981. "A Different Way to Record Classical Music." *Journal of the Audio Engineering Society* 29/5: 329–32.

Røde Microphones. 2011. "Interview with a Legend: Tony Faulkner." Available at: www.youtube.com/watch?v=8uCcFIyJJ-w.

Simmons, Greg. 2016. "Stereo Masterclass [Interview with Tony Faulkner]." *AudioTechnology*, 31 May. Available at: www.audiotechnology.com.au/wp/index.php/stereo-masterclass/.

CHAPTER 6

Tracking

CRITICAL LISTENING

Critical listening should occur continuously throughout the recording process, but during the tracking and editing/mixing stages of a project, engineers usually spend a great deal of time evaluating the sound quality of the analog information being captured and the digital information being processed in the DAW. The criteria listed below, based on a set of parameters recommended by the European Broadcasting Union for the appraisal of recorded material (1997: 13–15), provide guidelines for assessing various facets of audio signals:

Spatial Environment—appropriateness of the ambience for the sound sources: the duration of the room reverberation, balance between direct and indirect sound, depth perspective, homogeneity

Transparency—clarity of details: the relationship between reverberation and the intelligibility of instruments and/or words in sung text, the identification and differentiation of instruments and/or voices sounding together

Timbre—the accurate portrayal of the tonal characteristics of the sound sources (including the naturalness of their harmonics); distribution of frequencies across the audible spectrum

Loudness—the faithful capture of the dynamic range of the music: the relative strengths of the sound sources, variations of light and shade across or between phrases and larger sections of the music, the realism of the bloom after the onset of a sound (that is, the naturalness of the rate at which the sound increases over time)

Stereo Image—the realistic location of the sound sources in the stereo field; balance across the left/right plane; stability (sound sources remain in their designated positions)

Noise and Distortion—the perceptibility of any unwanted events, whether captured by the mics or incurred along the signal path

Taken as a whole, the criteria help direct the aural attention of engineers to essential characteristics of recordings that can either enhance or diminish the experience for listeners, and as the ability to listen critically increases, a greater number of these details can be ascertained on a first hearing. The subjective impressions that result from close listening, then, put the final shape on performances destined for transmission through loudspeakers.

The perception of room ambience, especially the reverberation time and the sonic characteristics of the reflections, help place soloists or ensembles in a natural sounding environment, and careful attention to this facet of the recording determines whether the reverberant qualities of the space need to be bolstered by artificial means, that is, by an algorithm chosen to match the room's ambience. Listeners must, of course, be able to identify and distinguish instruments and/ or voices within this wash of reverberation, so engineers like to ensure that the individual elements remain readily intelligible. The tone color of these sound sources, with their characteristic overtone patterns, gives a composition its distinctive sonority, and the natural spread of this information across the audible frequency spectrum can create a sense of openness in a recording. In addition, the dynamic range of the performance, complete with its variations of light and shade, engages the hearers' emotions directly.

When engineers place all these details in a realistic stereo image, the music may be experienced through a recording that replicates aural events as faithfully as technology allows. But such a finely crafted recording would remain fundamentally flawed if unwanted noise distracted listeners from the communication of the performers' emotion. Hence, after a master track has been compiled, and before any further editing or mixing takes place, engineers remove as much extraneous noise from the recording as possible.

SETTING LEVELS

Since all systems have a noise floor that can become audible when the gain is raised past a certain point, especially during post-production, an appropriate input level at the tracking stage allows recordists to optimize the signal-to-noise ratio, while leaving enough room for the avoidance of clipping (too low a level could cause the noise floor to become audible if the signal needs to be raised on final output, especially when preparing tracks for delivery to streaming services; see Part 3, Chapter 10 for further discussion). Before the advent of digital recording, analog consoles did not produce noticeable distortion until the input signal exceeded the equipment's nominal 0.0 dBu level by 20.0 or more decibels, which gave plenty of headroom for engineers to accommodate the highest transients. In digital systems, however, clipping occurs as soon as the signal crosses the 0.0 dBFS threshold, so recordists must leave enough decibel space below this point to prevent peak levels from causing distortion. As in analog equipment, approximately 20.0 dB of headroom (which means an average level falling roughly 20.0 dB below 0.0 dBFS) not only allows true-peak transients to be captured without clipping but also permits the noise floor of 24-bit systems to remain 90.0 to 100.0 dB below the average signal level. In this scenario, most transients probably would not rise above −10.0 dBFS, with extreme events staying at or under −5.0 to −6.0 dBFS.

To capture sound in an ideal way, then, recordists usually set their levels from the loudest passages, carefully adjusting the input controls to achieve maximum true-peak readings between 5.0 and 10.0 dB below full scale. With such a scheme in place, engineers can easily produce clip-free recordings with the optimal signal strength required for post-production processing (the historic practices of classical recordists are discussed in Part 3, Chapter 10, "Loudness and Meters").

ROOM AMBIENCE

During tracking, engineers have several options for dealing with room ambience, and depending on the nature of the space and the amount of natural reverberation desired, a pair of microphones can be situated closer to the source or farther away. Locations in the near field will mainly pick up direct sound, while positions in the diffuse field will only capture reverberation. The place in a room where direct and reverberant sound are equal in level is called the critical distance or reverberation radius (see Figure 6.1), and many recordists like to use this area as a starting point for determining the optimum balance between direct and reflected sound, bearing in mind that the more reverberant the room, the closer the critical distance will be to the sound source.

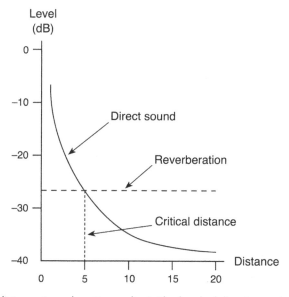

Figure 6.1 Critical distance (reverberation radius). The level of direct sound drops by 6.0 dB for every doubling of the distance in the free field (less of a drop in enclosed spaces), while the level of reverberated sound remains more or less constant everywhere in a room. The ratio of direct to reverberated sound is 1:1 at the critical distance.

An SPL meter can help engineers find the reverberation radius of any room. Either during a rehearsal or by situating a boom box at the performers' location (set between stations to produce white noise), recordists measure the SPL close to the source (approximately 30 centimeters or a foot away) and then double the distance and measure again, a point at which, according to the Inverse Square Law, the level will have decreased between 4.0 and 6.0 dB. After making note of the new level, they double the distance, and take another measurement, repeating this procedure until the SPL stops dropping. By moving back to the area where the level began to remain constant, they find the critical distance.

The reverberation radius, however, may not be an ideal location, as microphones situated there can make sources sound as though they are quite far away. Indeed, most engineers test positions within the critical distance to pinpoint what they call the sweet spot, a place that captures the best ratio of direct-to-reverberant sound. If, for example, the mics appear to be filled with so much detail that the hall ambience is excluded or if the sound is distant and muddy with no detail, the mics can gradually be moved in one direction or the other to achieve a more balanced perspective (depending, of course, on the capsule type employed—omnis can be placed closer than cardioids or bi-directionals; see Part 2, Chapter 4 for a discussion of distance factor).

If a single stereo pair of mics fails to produce satisfactory results (that is, the ratio of direct to reverberant sound does not sound natural), recordists often opt to use two pairs of microphones, one positioned closer to the sound source to provide clarity of detail and the other supplying ambience from the diffuse field. In post-production, engineers blend the signals to create the perception of depth, while maintaining an appropriate amount of detail. Some experimentation in the room is required so that the mixed signals do not contrast too much (for example, a dry, clear sound from the closer mics conflicting with the warmer reverberant character of the distant mics).

A third option involves placing mics close to the sound source to defeat room reflections completely. This approach requires the addition of artificial reverberation during post-production, and many engineers find this solution workable for concert halls with disappointing ambience (see the discussion of reverberation plugins in Part 3, Chapter 9). In deciding on what perspective to present to listeners in their home environment, recordists consider the size of the room they wish to emulate, from larger spaces to more intimate rooms.

REFERENCE

European Broadcasting Union (EBU). 1997. *Tech 3286: Assessment Methods for the Subjective Evaluation of the Quality of Sound Programme Material—Music*. Geneva, Switzerland: European Broadcasting Union.

PART 3

Post-Production

CHAPTER 7

EQ

DIGITAL FILTERS

In the early days of telephonic communication, long cable runs caused such a loss of high-frequency content that the quality of a signal's output was markedly inferior to the input. Electronics engineers compensated for this deficiency by employing a circuit to boost high frequencies. This circuit, called an equalizer, allowed the output to become roughly "equal" to the input.

Today, audio editors use equalization (EQ) to modify the frequency content of signals to enhance the aesthetic appeal of individual or combined tracks. Recordists often achieve these alterations through digital filters, devices normally classified according to type: pass, shelf, parametric, and graphic (I define the term "filter" as any device that changes the frequency spectrum of a signal by allowing some frequencies to pass, while attenuating others; consequently, filters change the balance between the various sine waves that constitute complex waveforms).

Pass filters let frequencies above or below a prescribed cut-off point pass through the filter at full level, but block all other frequencies in the spectrum. Engineers define the cut-off frequency as the point at which the response of the filter is 3.0 dB below the nominal level of the unaffected signal (see Figure 7.1).

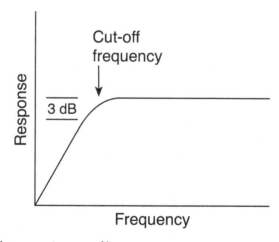

Figure 7.1 Cut-off frequency in a pass filter.

A high-pass filter (HPF) permits frequencies above the cut-off point to pass, while a low-pass filter (LPF) allows frequencies below the cut-off to pass. Editors may apply the attenuation gradually or more sharply, and the number of decibels of reduction across an octave span (usually in steps of 6.0 dB) determines the rate of attenuation or slope. A drop of 6.0 dB per octave produces a much gentler attenuation slope than a steeper descent of 12.0 dB per octave (see Figures 7.2 and 7.3 for examples of each). In band-pass filters, however, only frequencies within a defined range, or band, pass through the filter (see Figure 7.4).

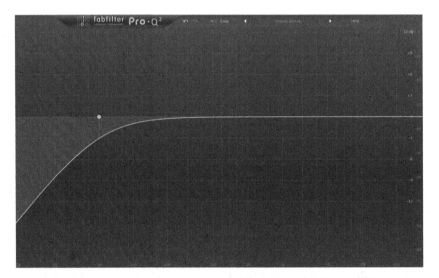

Figure 7.2 High-pass filter with a gentle slope (6.0 dB per octave; the dot indicates the cut-off point, and the shaded area shows the attenuated frequencies).

Source: Screenshot from *Pro-Q 2* used with the permission of FabFilter Software Instruments.

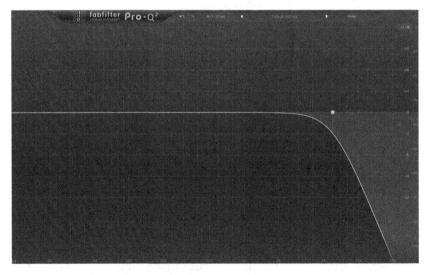

Figure 7.3 Low-pass filter with a steeper slope (12.0 dB per octave).

Source: Screenshot from *Pro-Q 2* used with the permission of FabFilter Software Instruments.

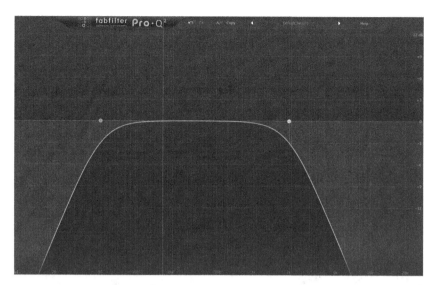

Figure 7.4 Band-pass filter with steep slopes (12.0 dB per octave; frequencies pass within the cut-off points of 50 Hz and 2 kHz).

Source: Screenshot from *Pro-Q 2* used with the permission of FabFilter Software Instruments.

Shelf filters operate somewhat differently. Unlike pass filters, which only cut frequencies, shelving filters can boost, as well as cut. They let all frequencies pass but uniformly boost or attenuate a specified range; that is, they increase or decrease the gain of the frequencies above or below a defined point (hence, these filters look like shelves; see Figure 7.5).

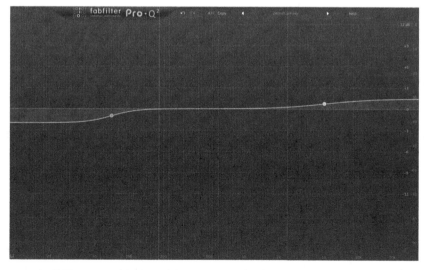

Figure 7.5 Shelf filters (gain change of 2.0 dB applied to lower frequencies and 1.5 dB to higher frequencies).

Source: Screenshot from *Pro-Q 2* used with the permission of FabFilter Software Instruments.

EQ   67

Parametric filters offer engineers a great deal of flexibility in shaping the frequency content of soundwaves, for beyond the three main adjustable parameters—frequency, gain, and Q—the digital domain provides further means for tailoring the spectrum. In the *Pro-Q 2* plugin, for example, FabFilter augments the three primary parameters with tools that enable the user to incorporate other types of filters. Figure 7.6 shows the basic operation of the *Pro-Q 2* filter. I have chosen two center frequencies (indicated by the dots) and have applied a gain change of –3.0 dB to one and +1.5 dB to the other.

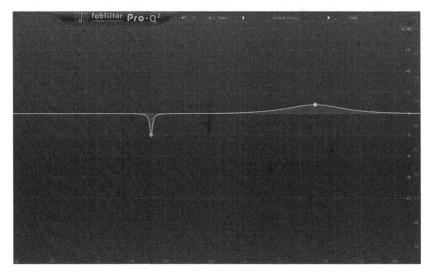

Figure 7.6 Basic use of a parametric filter (a narrow cut at 150 Hz and a wider boost at 4.0 kHz).

Source: Screenshot from *Pro-Q 2* used with the permission of FabFilter Software Instruments.

The Q or quality-factor setting establishes the degree to which these alterations affect the neighboring frequencies. Q refers to the frequency or band width of a cut or a boost, and audio engineers commonly define it as the region surrounding the center frequency within 3.0 dB of the maximum gain change applied to that frequency. Recordists have the option of either restricting the range of frequencies affected (a narrower Q setting of 15 on the left side of Figure 7.6) or spreading the change over a larger bandwidth for a subtler effect (a wider Q setting of 1 on the right side of Figure 7.6). Figure 7.7 shows the Q value for a wider (solid line) and a narrower (dotted line) bandwidth.

But the *Pro-Q 2* filter gives audio editors far more creative control over the frequency content of a track than the illustration of basic operation might suggest. Engineers can shape several frequency areas simultaneously through bell-style curves and can incorporate pass and shelf filters as well. In Figure 7.8, a high-pass filter removes rumble in the low end, while gentle boosts to two other areas add a touch of warmth around 100 Hz and extra presence in the 3.0 kHz range. To

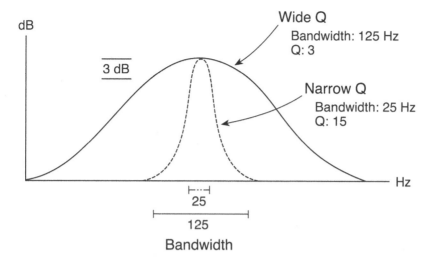

Figure 7.7 Bandwidth and Q.

lessen the brightness of the highest frequencies, a shelf filter attenuates the area above 10 kHz by a modest 1.5 dB. In addition, a narrow cut of 2.0 dB at 300 Hz helps remove some of the excess energy accumulated in that part of the spectrum.

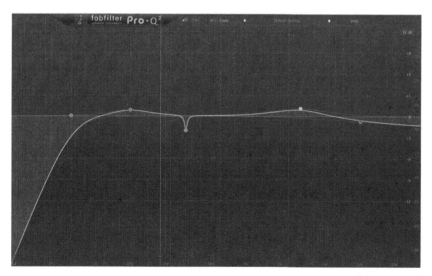

Figure 7.8 Further spectral shaping in FabFilter's *Pro-Q 2*.
Source: Screenshot from *Pro-Q 2* used with the permission of FabFilter Software Instruments.

Graphic filters consist of a number of separate bands, each with a bell-shaped response curve centering on a given frequency. Users generally adjust the gain with sliders, and in Figure 7.9, the 16 bands in Voxengo's *Marvel GEQ* alter the frequency balance by up to 12.0 dB between 20 Hz and 20 kHz.

EQ   **69**

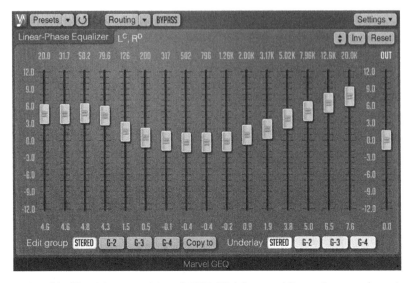

Figure 7.9 Graphic filter, Voxengo, *Marvel GEQ* ("Brighter and Brassy" preset shown).
Source: Used with the permission of Voxengo.

COMMON PRACTICES (A PLACE TO START AS LISTENING SKILLS DEVELOP)

1. High-pass filter

   - helps to remove undesirable low-frequency information ("muddiness," "rumble," etc.)
   - 40–100 Hz

2. Shelf filter for higher frequencies

   - the impression of "sheen" can be attained by gently boosting the frequencies above 8–10 kHz

3. Sweeping

   - offending frequencies can be difficult to hear, and sweeping the mid-range with a heightened bell curve can help find them
   - often 800 Hz—1 kHz, or thereabouts
   - cut slightly to lessen their effect

4. Brightness

   - 2–5 kHz
   - if desired, small boosts in this range can give a bit more brightness
   - human hearing is the most sensitive to sounds in the 2–5 kHz range

5. Other low-frequency areas

   - listen in the 100–300 Hz range for any further "muddiness" that might negatively affect the frequency balance of the track

Note: Human hearing is not equally sensitive to all frequencies, and during the mixing process, playback level affects how listeners perceive frequency. In the 1930s, the researchers Fletcher and Munson studied the way people respond to frequencies across the entire range of human hearing. Over the years, refinements have been made to their findings, and a chart known as the equal loudness curves (contours) shows how humans typically respond to frequencies at different levels of loudness (see Figure 7.10). The contours tend to flatten out around 80.0 dB, a level close to the normal listening volume of 85.0 dB, so the balance audio engineers achieve across the frequency spectrum when monitoring at 80.0 to 85.0 dB should be more or less correct. At lower levels, bass frequencies sound weaker, while higher levels seem to exaggerate the bass. The usual listening practices of the target audience, then, should determine the level for mixing.

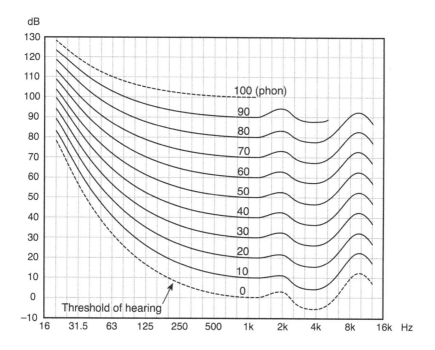

Figure 7.10 Equal loudness curves (contours).

REFERENCES

FabFilter. n.d. *Pro-Q 2*. Amsterdam, The Netherlands.
Voxengo. n.d. *Marvel GEQ*. Syktyvkar, Russia.

EQ   71

CHAPTER 8

Control of Dynamic Range

COMPRESSORS

First invented for use in "live" radio broadcasts as a way of preventing the level of an announcer's voice from suddenly overloading the signal chain, compressors automate the process of controlling gain at moments when spikes would be impossible for audio engineers to predict and attenuate quickly enough. In the field of music production, recordists have also chosen to restrict the dynamic range of signals, particularly in pop/rock, where uniform levels have become the norm, and in classical genres, even though engineers prefer to retain the full dynamic range of the performances they capture, the control of dynamics has its usefulness in editing, mixing, and delivery, especially during the preparation of tracks for streaming.

Before the invention of compressors, engineers relied on faders to reduce dynamics manually, and they often "rode" the faders in anticipation of excessive peaks. Modern compressors, however, can react instantaneously to fluctuations in level, and they easily balance the dynamics from note to note by decreasing the differences in amplitude between the loud and soft parts. With the advent of digital signal processing, the compressor has evolved into an important tool for helping audio editors shape the dynamic range of individual tracks or complete recordings. These plugins process the signal in a side chain that compares a measured signal to a reference level set by the user. Compressors consist of three main stages: level detection (either peak or RMS), gain control, and make-up gain (see Figure 8.1). The gain-control stage contains several user-adjustable features, such as threshold, ratio, attack, release, knee, and hold. After the compression process has been completed, engineers often apply make-up gain to amplify the output of the compressor so that the perceived loudness of the overall signal appears greater.

The *Peak/RMS* switch in a compressor allows users to choose the way in which the plugin detects the level of a signal. When an engineer selects "peak," the compressor responds to the peak values of the signal, but in the RMS setting, the compressor reacts to the average loudness of the signal over a specified period (see Figure 8.2 for a comparison of peak, RMS, and average measurements). The letters RMS abbreviate the expression "root mean square" (the square root of the mean of the squares, which involves squaring the signal mathematically and then averaging it over a period of time; White & Louie 2005: 335), and this method of measuring sound approximates the way humans perceive loudness.

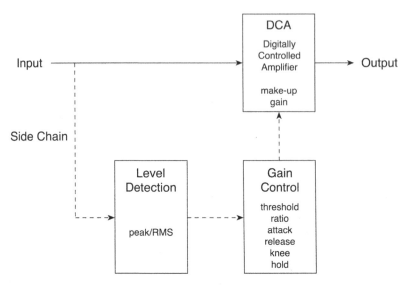

Figure 8.1 The main stages employed in compressors.

In fact, some engineers believe that both instruments and voices sound better when compressors respond to averaged loudness instead of peaks, because the outcome more closely aligns to our perception of loudness (for a fuller discussion of loudness, see Part 3, Chapter 10).

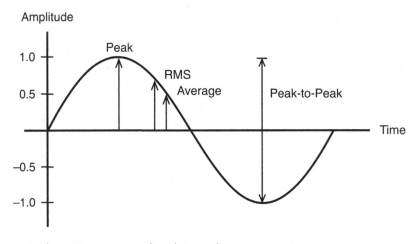

Figure 8.2 Peak, RMS, average, and peak-to-peak measurements.

Threshold sets the level (in dB) at which the compressor will begin to reduce the amplitude of a signal. Any level below the threshold does not activate the compressor, but when the signal rises above the threshold, the compressor automatically lowers the level in a manner akin to someone moving a fader. The diagrams in Figure 8.3 illustrate how much signal a compressor can affect (peaks are shown at –2.0 dB and thresholds at –6.0 dB and –10.0 dB).

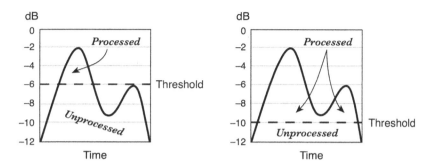

Figure 8.3 Threshold (two different amounts of treatment applied to a transient spike).

Ratio determines the amount of gain reduction that occurs when a signal exceeds the threshold, and plugin designers think of a compressor's output level as a comparison between input and output. They express this comparison in terms of a decibel ratio—change in the input level:change in the output level (read the colon as "to"). For example, if the input level increases by 2.0 dB and the output level rises by only 1.0 dB, the compression ratio is 2:1. At a ratio of 4:1, a 4.0 dB increase in input level also produces an output rise of 1.0 dB. Thus, the higher the ratio, the more the compressor reduces the dynamic range of the signal. Figure 8.4 demonstrates how ratio works in relation to the threshold.

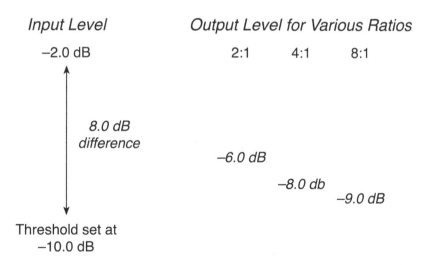

Figure 8.4 Ratio.

2:1 ratio—every 2.0 dB above the threshold at input is reduced to 1.0 dB above the threshold at output; the level difference of 8.0 dB between input and threshold contains four 2.0 dB units, thus the 2:1 ratio lowers the input from 8.0 dB to 4.0 dB above the threshold, so that the signal emerges from the compressor at −6.0 dB.

4:1 ratio—every 4.0 dB above the threshold at input is reduced to 1.0 dB above the threshold at output; the level difference of 8.0 dB between input and threshold contains two 4.0 dB units, thus the 4:1 ratio lowers the input from 8.0 dB to 2.0 dB above the threshold, so that the signal emerges from the compressor at −8.0 dB.

8:1 ratio—every 8.0 dB above the threshold at input is reduced to 1.0 dB above the threshold at output; the level difference of 8.0 dB between input and threshold contains one 8.0 dB unit, thus the 8:1 ratio lowers the input from 8.0 dB to 1.0 dB above the threshold, so that the signal emerges from the compressor at −9.0 dB.

This information may also be shown in graph form (see Figure 8.5), and as stated above, the greater the ratio, the more the compressor attenuates the signal. At ratios beyond 10:1, the plugin effectively becomes a brickwall limiter, as the output level does not exceed the threshold (see Figure 8.6; limiters are discussed in the next section).

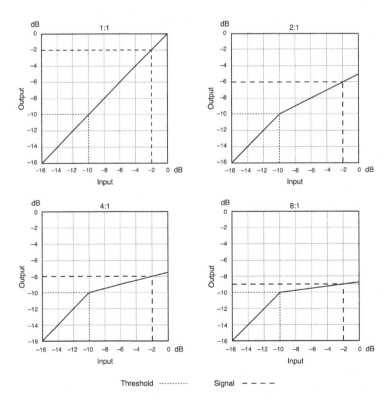

Figure 8.5 Various ratios shown in graph form.

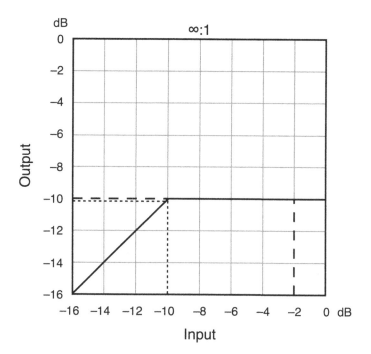

Figure 8.6 Ratio of ∞:1.

Attack and *release* allow users to shape the dynamic response of the compressor. *Attack* controls the length of time between the moment an audio signal exceeds the threshold and the start of gain reduction, while *release* determines how quickly the compressor returns the attenuated signal back to its original level.

The initial onset of sound is an important factor in defining the character of an instrument or voice, so longer attack times (20–50 ms) often provide a more natural representation of the source. Similarly, longer release times (c.135 ms) make the transition back to the original level less obvious. Short release times can create a "pumping" effect on certain signals (that is, one hears the volume repeatedly increasing at too quick a rate). Figure 8.7 shows the effect of attack, gain reduction, and release on an idealized waveform (the middle graph schematically depicts the action of the compressor).

The *knee* function within a compressor softens the onset of gain reduction somewhat, for it allows compression to be introduced gradually across the threshold (soft knee), instead of suddenly (hard knee; see Figure 8.8). In addition, some plugins feature a *hold* control, and this parameter lets users decide how long the compressor "holds" the gain reduction before the release begins.

Since the process of compression reduces the dynamic level of signals exceeding the threshold, the perceived loudness of the signal leaving the gain-modification stage of the side chain can seem quieter than the signal that entered the compressor. Engineers compensate for this apparent discrepancy through the

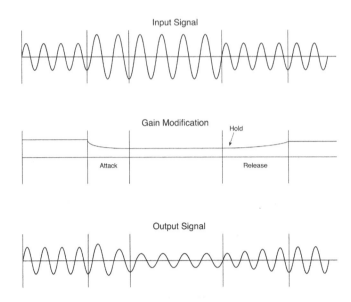

Figure 8.7 Schematic of a compressor's action. Adapted from FabFilter n.d.: 6.
Source: Used with the permission of FabFilter Software Instruments.

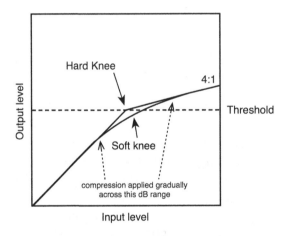

Figure 8.8 Knee function of a compressor.

final *gain* control (also known as *make-up gain* or *output*). By uniformly boosting the output of the gain-modification stage with whatever number of decibels seems appropriate, engineers regain lost ground, so to speak. This amplification takes place after the side chain has processed the signal (see the block diagram in Figure 8.1 for a visual representation of this), and the increased gain affects the unaltered portion of the signal below the threshold, as well as the compressed portion above the threshold. Figure 8.9 shows the controls available in Sonnox's plugin *Oxford Dynamics*.

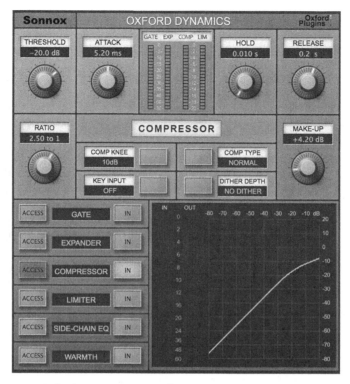

Figure 8.9 Sonnox, *Oxford Dynamics*, controls.

Source: Used with the permission of Sonnox.

LIMITERS

As mentioned above, limiters are specialized compressors that allow audio below a specified decibel level to pass freely, while attenuating any peaks which cross a user-defined threshold. They have a ratio of at least 10:1, as well as a fast attack time, but if software developers give them a ratio considerably higher than 10:1 and a very short attack time, the plugins become brickwall limiters, programs that ensure no peaks cross the threshold.

Nugen Audio designed their true-peak limiter *ISL 2* as a brickwall device that not only measures inter-sample peaks but also permits engineers to set the maximum decibel level signals may reach (limiting audio through true-peak measurements makes certain that downstream codecs, such as mp3 or AAC, will not add distortion to the signal; codecs and the concept of true peak are discussed in Part 3, Chapter 10). The *Input* section of the plugin (the left side of Figure 8.10) indicates the true-peak level of the audio entering the limiter, and engineers may adjust the amplitude of the incoming signal through the *Input Gain* box below the left meter. Users select a target for the true-peak limit of the signal either through the TP_{Lim} box below the right meter or by dragging the arrows on either side of the input meters up and down.

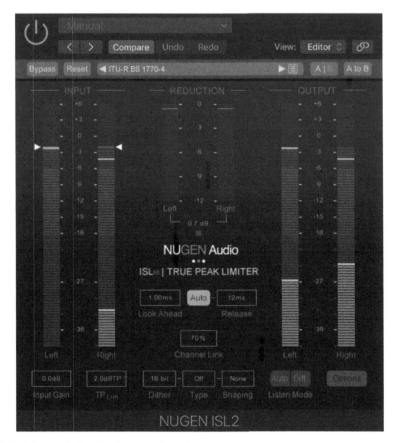

Figure 8.10 Nugen Audio, *ISL 2*.

Source: Used with the permission of Nugen Audio.

The center section of the graphical user interface (GUI) displays the amount of gain reduction that has been applied to each channel (calculated in dBTP), and by clicking on the small *H* below the gain-reduction meters, a scrolling history graph shows the input level, as well as the gain reduction, the limiter has imposed (see the upper central portion of Figure 8.11). Two additional meters underneath the history graph measure *steering* ("S") and *ducking* ("D"). When limiters are employed independently on the two stereo channels and only one side triggers the device, the gain reduction is applied to one channel but not the other. The resulting gain imbalance creates the perception that the center of the stereo image has "steered" away from its original position, and the meter provides an indication of the degree of steering that has occurred. Conversely, when the two channels are fully linked and only one side triggers the single limiter applied to both of them, the plugin reduces the left and right sides equally. In this situation, the second channel receives gain reduction unnecessarily, and the meter shows the extent to which the processing has inadvertently introduced the effect known as "ducking" (one signal dropped below another).

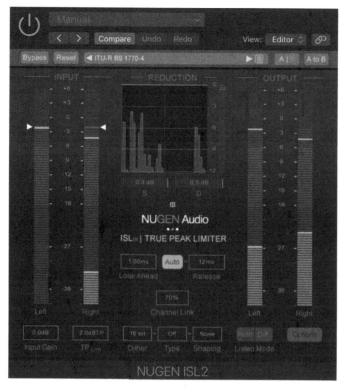

Figure 8.11 *ISL 2*—history graph, steering, and ducking.

Source: Used with the permission of Nugen Audio.

The lower portion of the center section contains four additional features: look ahead, auto release, release, and channel link. The *Look Ahead* box allows engineers to specify how much time the limiter should be given to react to the entering signal in advance of processing the audio (smaller values preserve the transients better than larger ones), and the *Release* function establishes how quickly the limiter returns to no gain reduction. When users activate the *Auto* button, the plugin analyzes the incoming audio signal for low-frequency information and increases the "hold" parameter of the limiter to provide enough time for a complete wavelength to pass through the device before the plugin commences the release stage of the process. The *Channel Link* box determines the amount of independence the individual limiters have. At 0%, the stereo channels are completely isolated from one another, and at 100%, the channels are fully linked, so that both the left and right signals receive the gain reduction applied to the side with the highest input level. Values between the two extremes control the degree of independence (at 50%, the steering and ducking values are equal).

The right section of the GUI contains the output meters, as well as a feature called *Listen Mode*. The meters show the output level in dBTP, and *Listen Mode* lets engineers compare the input and output. The *Auto* button adjusts the output in accordance with the amount of *Input Gain* selected when the signal entered the plugin, and this innovative component allows users to hear the limiter's effect on the signal without the initial level adjustment present at the output. The *Diff* (difference) button lets engineers listen to just the gain reduction that has been imposed on the audio (that is, recordists hear only the difference between the input and the output).

ISL 2 users may also opt to add final-stage dither through the *Dither, Type,* and *Shaping* boxes below the *Channel Link* button. When engaged, Nugen's dithering algorithm compensates for any increase in gain caused by dithering, so that the true-peak level leaving the limiter does not exceed the desired maximum. The *Dither* box sets the target bit depth (16, 18, or 20), while the *Type* box lets engineers choose between three forms of white noise, as defined by the probability density function (PDF) used to generate the noise—rectangular (RPDF), triangular (TPDF), and Gaussian (GPDF). The *Shaping* function specifies the nature of the noise shaping, which ranges from none to three procedures for shifting the quantization error into less audible frequencies. *Nyq* reduces the signal's overall noise floor, but increases the noise above 17 kHz, whereas *Psy 1* and *Psy 2* follow established psychoacoustic principles to reduce noise in the frequency areas people hear the best.

DYNAMIC EQ

Compression can also be applied in conjunction with an EQ filter, and the German company Brainworx has developed software it calls the *bx_dynEQ* (illustrated in Figure 8.12). The accompanying manual defines this dynamic EQ as "a filter that is not limited to being set to a specific gain level, but which changes its gain settings dynamically—following the dynamics of a certain trigger signal" (Brainworx 2018b: 4). Engineers use the *bx_dynEQ*, then, to isolate and correct frequency problems that arise when a signal crosses a specified threshold, so that editors may remove, for example, harshness heard only at or above a particular dB level. Traditional static EQs would be unsuitable for this kind of intermittent correction, because they would continuously cut the frequency area, even when the problem was not present. Brainworx includes a number of filter types—parametric (peak band), high/low pass, band pass, and high/low shelf—to assist engineers in applying gain reduction to user-defined frequency bands.

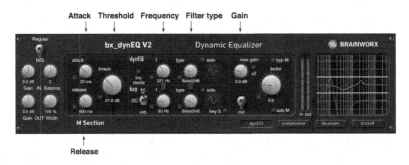

Figure 8.12 Brainworx, *bx_dynEQ*.

Source: Used with the permission of Brainworx Audio.

Sonnox takes a somewhat different approach to the design of the GUI for their *Oxford Dynamic EQ*. Instead of replicating the control knobs of a hardware unit, as Brainworx has done, Sonnox allows users to adjust certain parameters through a graphical interface similar to those found in parametric EQ plugins (see Figure 8.13). Like other dynamic EQs, the Oxford software filters the input signal before sending that altered signal to a dynamics processor, and Sonnox provides three types of filter (low shelf, bell, and high shelf) that may be applied to each of five distinct frequency bands (numbered 1 to 5 in Figure 8.13). By dragging the control points to the left or right, the user may adjust either the center (bell) or the corner (shelf) frequency, and an up or down mouse motion determines the target gain for the band. If a bell filter has been chosen, holding the option key (Mac) while moving the mouse up and down will widen or narrow the Q. The numerical values associated with the chosen settings are shown below the graph on a line labeled EQ, and at the end of this line, the user may set the offset, by which Sonnox means the resting or static gain of the equalizer band. In fact, the plugin has been designed to constrain the amount of gain change between the offset and target settings so that, in Sonnox's words, "over-processing" cannot occur (Sonnox 2017a: 2). Moreover, when engineers click on the headphone icon in the EQ line, they may listen to just the frequency range that will be processed. A Fast Fourier transform (FFT) display shows the frequency spectrum of the output signal.

Engineers control the dynamics processing they wish to apply to the five EQ bands through the group of knobs displayed at the bottom of the screenshot. On the left side of the panel, the *Detect* column allows users to choose the way the plugin identifies the level of a signal (overall peak level or sudden increases at the

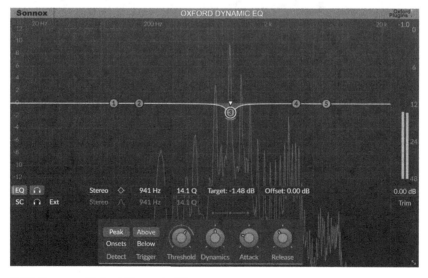

Figure 8.13 Sonnox, *Oxford Dynamic EQ*.
Source: Used with the permission of Sonnox.

onset of transients). The *Trigger* section establishes whether compression will be applied when the signal rises above or falls below the threshold, and engineers adjust the *Threshold* knob to set the level at which the processing will be activated. The *Dynamics* controller alters how the gain is applied. At high settings, the gain is "more likely to reach the target level," while positioning the knob at 0% causes the band to remain at the offset gain (Sonnox 2018: Tooltip). The *Attack* parameter determines how quickly the band approaches the target level, and the *Release* knob sets how slowly the band will return to the offset level.

Engineers adjust the overall output of the plugin through the *Trim* control at the right of Figure 8.13. The gain can be set to avoid any clipping that might occur as a result of the processing, or the control can be used to match the levels of the input and output signals.

DE-ESSERS

When audio engineers record a vocalist through a closely placed microphone (the near field), they run the risk of exaggerating the sibilance associated with the consonants "s," "sh," "ch," and "z." Any sort of unnatural emphasis on the high-frequency content of sibilant sounds quickly tires listeners, and if engineers do not reduce the energy of the bands containing the offensive frequencies, further processing, particularly compression, limiting, or equalization (for brightness), can intensify the problem.

Some audio editors choose to lessen the effect of harsh consonants manually, either through track automation (see Figure 8.14) or by cutting out the sibilant moments and placing them on a separate track reduced in level (see Figure 8.15).

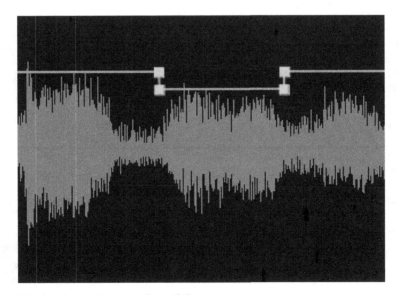

Figure 8.14 Fader automation to reduce sibilance.

Source: Screenshot from *Mixbus* used with the permission of Harrison Consoles.

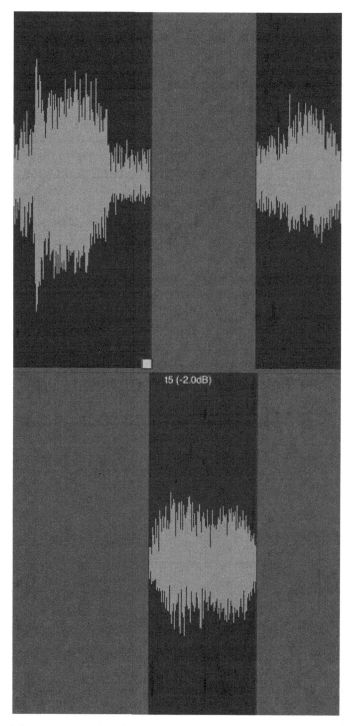

t5 (-2.0dB)

Figure 8.15 Sibilant moments placed on a separate track, with gain reduced by 2.0 dB.
Source: Screenshot from *Mixbus* used with the permission of Harrison Consoles.

But beyond reducing the level of the sibilance, editors also frequently apply "notched" equalization within the 4–10 kHz range to help them ease the stridency of the problematic areas. All these techniques alleviate sibilance, but since many people find the manual approach far too time consuming, plugin designers have developed specialized compressors, known as de-essers, to automate the process.

Like all compressors, de-essers work on the principle of side chaining. In the simplest theoretical model, the de-esser divides the input and sends it along two paths. One path retains the original vocal track (the input) and goes to the compressor. The other, which consists of a modified version of the input signal, travels along a side chain to the level detector. This side signal receives a boost of around 12.0 dB (or more) in the narrow high-frequency band that contains the sibilance, and the de-esser uses the heightened energy of this band to trigger the compressor to attenuate those frequencies in the original track that the side chain had boosted. In other words, this design applies compression exclusively to the unmodified input. Listeners never hear the filtered signal sent along the side chain, for that pathway functions only to trigger the compressor. These types of systems decrease the harshness of the sibilance, while allowing the remaining components of the vocal's frequency spectrum to pass through the device unaltered.

The Sonnox *SuprEsser* employs similar principles and offers audio editors considerable flexibility in treating excessive sibilance (see Figure 8.16). An explanation of its main features will aid our understanding of the de-essing process.

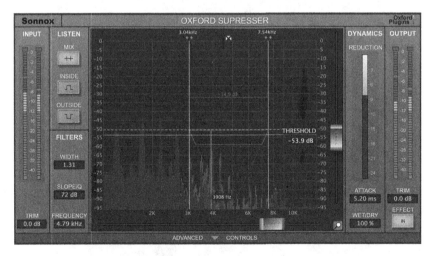

Figure 8.16 Sonnox, *Oxford SuprEsser*.
Source: Used with the permission of Sonnox.

The software engineers at Sonnox have designed the *SuprEsser* as a compressor linked to two filters (narrow band-pass and an equally narrow band-reject). The plugin sends the input to both filters, but only the content of the signal leaving the band-pass filter goes to the compressor for attenuation. In other words, the band-pass filter isolates the problematic audio in the frequency range specified by the user. The signal emerging from the band-reject filter contains the remaining frequency components of the input (that is, the slim band of isolated frequencies sent to the compressor has been removed completely), and when the crossover block (the *Listen* section) mixes the two signal paths together, the original signal, minus the excessive sibilance, results.

Three listening modes allow users to hear the operation of the plugin from differing perspectives. The *Inside* button solos the output of the band-pass filter, while the *Outside* button reveals the output of the band-reject filter. The *Mix* button blends the two in a 50:50 ratio.

A set of graphical controls lets editors finely adjust a number of parameters (see Figure 8.17). The upper and lower limits of the band-pass filter can easily be set by dragging the vertical lines at numbers 1 and 3 in one direction or the other. After the user has isolated problematic frequencies as narrowly as possible (by dragging the lines defining the upper and lower limits of the band-pass filter), the horizontal line at 9 indicates the peak level of the problematic sound in a peak-hold manner. The editor may now set the threshold level (the horizontal line at 4) to the desired point below the peak level indicator (9).

In the default setting, the plugin automatically adjusts the threshold so the device produces the same amount of gain reduction when transient peaks occur in both quieter and louder moments. The gain-reduction meter (labeled

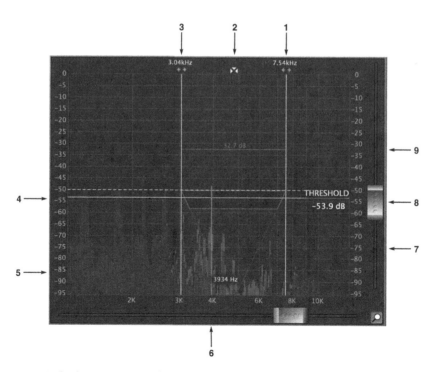

Figure 8.17 *Oxford SuprEsser*, graphical controls.

Source: Used with the permission of Sonnox.

Reduction in Figure 8.16) shows the extent of the gain reduction achieved with the setting, and as the automated threshold starts to follow the general level, the peak reductions should remain the same (that is, the attenuation meter should continue to display identical amounts of peak reduction).

In addition to these controls, *SuprEsser* represents the input signal as an FFT waveform (5), with the shaded peaks directly above it (7) illustrating the effect of gain reduction (the graph also provides the latter information in real time through the shaded area at 8, just below the threshold line). Beyond this, the plugin uses the vertical line at 6 to show which frequency within the band-pass window has the greatest energy. Clicking on this line centers the band-pass filter on that frequency. Furthermore, the user may drag the center frequency of the band window up and down the scale by clicking and dragging the icon (2) at the top of the graph; however, clicking any point on the graph also centers the band filter around the chosen frequency.

REFERENCES

Brainworx. 2018a. *bx_dynEQ V2*. Leichlingen, Germany.
Brainworx. 2018b. *Manual, bx_dynEQ V2*. Leichlingen, Germany.
FabFilter. n.d. *Manual, Pro-C*. Amsterdam, The Netherlands.
Nugen Audio. 2017a. *ISL 2*. Leeds, UK.
Nugen Audio. 2017b. *ISL 2, Operation Manual*. Leeds, UK.
Sonnox. 2017a. *Oxford Dynamic EQ, Manual*. Finstock, UK.
Sonnox. 2017b. *Oxford Dynamics*. Finstock, UK.
Sonnox. 2017c. *Oxford SuprEsser*. Finstock, UK.
Sonnox. 2018. *Oxford Dynamic EQ*. Finstock, UK.
Sonnox. n.d. a. *Oxford Dynamics, User Guide*. Finstock, UK.
Sonnox. n.d. b. *Oxford SuprEsser, User Guide*. Finstock, UK.
White, Glenn D. and Gary J. Louie. 2005. *The Audio Dictionary*. 3rd ed. Seattle, WA: University of Washington Press.

CHAPTER 9

Reverberation

In the world of recording, many engineers, particularly classical recordists, like to capture sound the way it occurs naturally in a reverberant space, while others, especially in the pop/rock field, choose to record direct sound with as little influence from the room as possible. But because listeners prefer to hear music in a naturally ambient setting, pop/rock recordists regularly add synthetic reverberation during the mixing stage. Consequently, virtually all recordings include both direct and reverberated sound so that the music sounds pleasant in all listening environments, for as Alexander Case (2007: 263) suggests, "the most important reverberation exists within the recording, not the playback space."

With the advent of software designed to mimic the behavior of sound in various types of spaces, audio engineers have the ability to enhance the character of natural room ambience through a wide range of algorithms for simulating reverberation, and these plugins are particularly useful when recording in rooms with less than ideal acoustics. The software follows one of two main methods of modeling, reflection simulation and convolution. Reflection simulation replicates the properties of sound propagating in a room, while convolution uses recordings of the impulse responses of specific rooms to reproduce the characteristics of those spaces.

DIGITAL REFLECTION SIMULATION

Plugins of the reflection-simulation type employ the main characteristics of room sound, early and late reflections, to generate spatial effects. By combining these two attributes in various ways, users can create an aural image of reverberant space (including the texture and timbre of the reflections). In fact, plugins traditionally divide room reverberation into its constituent parts and provide controls for direct sound, pre-delay, early reflections, late reflections (also called the reverb tail), decay time, room size and shape, density, diffusion, and equalization. Figure 9.1 shows the options available in Sonnox's *Oxford Reverb*.

Engineers often adjust the amount of *direct sound* (also called dry sound) they wish to include in the reverb signal through a slider or knob that varies the ratio between wet/reverberated or processed sound and dry or unprocessed sound (see the *Output* section of *Oxford Reverb* in the lower right corner of Figure 9.1).

Pre-delay refers to the time gap between the arrivals of the first wavefront (direct sound) and the initial reflection. The length of this gap varies according to the size of room and the listening position, and acousticians generally agree that smaller intimate halls have pre-delays of 15–18 ms or less. Larger rooms have longer pre-delays, as it takes more time for the reflected sound to reach hearers in spaces with widely separated boundaries. The pre-delay function of a plugin permits users to define the point at which the reflections begin to interact with

Figure 9.1 Sonnox, *Oxford Reverb*.

Source: Used with the permission of Sonnox.

the direct sound, and settings between 10 and 30 ms replicate the natural characteristics of smaller and larger spaces. Many plugins offer engineers further tools for sculpting the character of the early reflections, and Sonnox's *Oxford Reverb*, for example, not only lets users set both the shape and size of the simulated room but also allows recordists to assign the listening position within that space (see the *Early Reflections* pane in Figure 9.1).

But beyond these features, Sonnox includes early-reflection controls for width, taper, feed along, feedback, and absorption. *Width* determines the stereo separation of the reflections, while *Taper* affects the loudness of the reflections in relation to the distance they travel (the more the sound bounces around the room, the more the levels are reduced). *Feed Along* allows users to specify the amount of reinjection of the distributed sound, and this enables them to control the reverb's density. *Feedback*, however, adjusts the proportion of reflections that will be recirculated within the simulated environment. By increasing the feedback, engineers can lengthen the duration of the reflections and increase the effect of room-mode frequencies (such as "boominess"). *Absorption* models the amount of high-frequency reduction that can occur in a room as the soundwaves interact with various surfaces. In other words, with this control, the user can mimic the nature of the reflective surfaces (ranging from the greater absorption of soft furnishings to the lesser absorption of harder walls).

Plugins routinely offer users a number of tools to shape the late reflections or reverb tail, and *Oxford Reverb* has controls available for reverb time, overall size, dispersion, phase difference, phase modulation, absorption, and diversity (see the *Reverb Tail* pane in Figure 9.1). *Reverb Time* sets the length of time in seconds that it will take for the tail to fade to silence, and *Overall Size* creates the aural image of space through the size of the delays within the tail, larger settings providing the greatest sense of space but a slower buildup in the density of the reflections. *Dispersion* manages the rate at which the reflections build over time and gives the engineer a degree of control over the complexity and texture of the reverberation. *Phase Difference* allows users to manipulate the rate at which phase disparity grows between right and left channels, with greater settings causing a "widening and deepening of the stereo sound field" (Sonnox n.d.: 22). *Phase Modulation* slightly varies the reverb character over time to enhance the "realism and presence" of the reverberation (Sonnox n.d.: 23), and *Absorption* mimics the effect reflective surfaces have on higher frequencies. *Diversity* alters the width of the reverb's stereo image. The lowest position centers the reverberation quite narrowly, while the maximum position spreads it equally across the sound stage.

A number of plugins have an EQ section built in to them, so that users may more closely emulate the natural frequency response of physical spaces (see Figure 9.2 for an illustration of the equalizer in *Oxford Reverb*). Because rooms absorb high frequencies more easily than low frequencies, reverberation normally contains a reduced amount of high-frequency content. To make the reverb sound as realistic as possible, then, editors often use the EQ section to lessen the prominence of higher frequencies, especially if a dry signal might sound unnaturally bright after artificial reverberation has been added. Greater warmth may also be created through a gentle boost to lower frequencies, but if the low-end content in a track already makes the reverb sound too "boomy," judiciously applied EQ can mitigate the effect of the undesirable frequencies.

Figure 9.2 *Oxford Reverb*, EQ section.

Source: Used with the permission of Sonnox.

Exponential Audio takes quite a different approach to designing reverberation plugins. The GUI of *Nimbus* (see Figure 9.3) has three main panes, with the narrow middle one containing dials for modifying the mix of wet and dry signal, pre-delay time, reverb (decay) time, and overall level (trim), while the knobs in the right pane set the reverb size, crossover frequency, low-mid balance, damping frequency, damping factor, and width, as well as reverb delay time, tail suppression, and tail recovery (*Nimbus* launches with the *Tail* pane open). The left pane includes a meter area (EQ curves for the input and output filters and a "live" display of the output signal's frequency content) and controls for both the gain and EQ of the early reflections and the reverb tail.

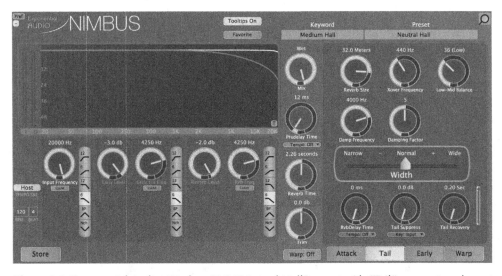

Figure 9.3 Exponential Audio, *Nimbus* GUI ("Neutral Hall" preset with "Tail" parameters shown).
Source: Used with the permission of Exponential Audio.

Within these adjustable parameters, an appropriate amount of *Pre-delay* helps increase the clarity and intelligibility of the signal, whereas a carefully chosen *Reverb Time* (decay) gives listeners a sense of what Exponential Audio calls the "reflectivity" of the space (for example, harder surfaces facilitate reflections and provide the means for the reverberation to last longer; Exponential 2018b: 39). By changing the *Reverb Size* (see Figure 9.4 for an illustration of the *Tail* pane), that is, the dimensions of the room, users can accelerate or decelerate the buildup of the reverb (in larger spaces, this accumulation of reflections occurs more slowly, which makes the reverb sound less dense). The *Crossover Frequency* knob lets engineers determine the point at which lower frequencies will be divided from higher ones (in large spaces, reverberation lasts longer in lower frequencies), and the *Low-Mid Balance* dial adjusts the relative reverb time below and above the crossover frequency. In the center position (the knob at 50), both the lower and upper frequencies will have identical reverb times, while smaller values (under 50) create longer decay times for the lower frequencies and larger values (above 50) do the same for upper frequencies. *Damping* controls the way the highest frequencies die away or roll off (air absorption and room treatment cause them to recede more quickly than lower frequencies), and the *Damping Frequency* knob sets the frequency above which damping takes place. The *Damping Factor* dial, on the other hand, determines the strength of the roll-off. The middle range of the knob approximates what Exponential Audio calls "normal" damping (Exponential 2018b: 42), with smaller values producing a darker sound, as the highs die away very quickly, and larger values providing a brighter sound, as the higher frequencies (at least those above the damping frequency) roll off slowly. The *Width* control applies only to the tail of the reverb (early reflections are not affected by it), and a wider setting will "open up" the space, while a narrower tail will "focus" the sound "more tightly" (Exponential 2018b: 42). *Reverb Delay Time* lets engineers introduce a degree of separation between the early reflections and the tail, which may help add "clarity" to a mix (Exponential 2018b: 41). The *Tail Suppress* feature lowers the level of the late reflections when a loud input signal might over-trigger the reverb, and the *Tail Recovery* dial sets the length of time it takes for the reverb to return to its original level after tail suppression ends. Short recovery values combined with subtle suppression levels can "make suppression effective without being obvious" (Exponential 2018b: 41).

Figure 9.4 *Nimbus*, tail pane.

Source: Used with the permission of Exponential Audio.

The *Output* pane (see Figure 9.5) contains separate dials for controlling the level and EQ of the early and late reflections, as well as a selection of filters for both the input signal and the reflections (the cut-off frequencies are set through the three *Frequency* dials). Six filter types allow engineers to craft the tonal character of the signal entering and exiting the plugin.

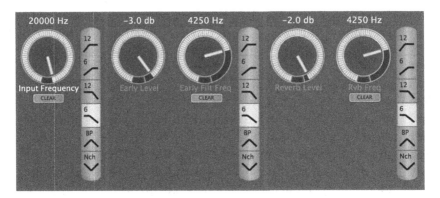

Figure 9.5 *Nimbus*, output pane.

Source: Used with the permission of Exponential Audio.

Two further sets of controls, available through the *Attack* and *Early* subpage buttons below the right pane, govern reverb attack and early reflections, while the *Warp* parameters "condition" the input signal (Exponential 2018b: 26). At the top of the *Attack* pane (see Figure 9.6), engineers may select between three general types of reverb—plate, chamber, and hall—and set the diffuser size, as well as the degree of diffusion present in the reverb. Plates have the densest reflections, with chambers also being somewhat dense, but in halls the reflections are sparser. Through the *Diffuser Size* knob, engineers model the dimensions of the irregularities on or near the reflective surfaces (that is, the imaginary objects in front of the surfaces or the materials that cover the walls, floor, and ceiling), and the *Diffusion* dial controls the overall amount of irregularity or diffusion in the space. Because the nature of a room's surfaces determines the way soundwaves bounce off the objects they strike, reflections may vary from a single bounce to multiple ones with small time delays between them (Exponential 2018b: 40).

In the next row of knobs, *Envelope Attack* controls the way the signal enters the plugin, and the vertical bars in the graph above the dial show an approximation of the number of reflections. With smaller attack values, the early audio energy is stronger, but with larger values, the later energy is stronger. A medium setting ensures that the plugin distributes the audio energy evenly. *Envelope Time* adjusts how long it takes for the signal to enter the plugin. In a short envelope, the signal injects into the plugin more quickly (over a shorter time frame), and a longer envelope injects the signal more slowly (larger values make the reverb "speak" more gradually; Exponential 2018a: Tooltip). The *Envelope Slope* controller consists of a low-pass filter that models air absorption (coloration of the vertical bars in the graph illustrates the effect), and various settings of the knob change the way the filter affects the signal as it enters the plugin. A gradual slope filters the later energy quite strongly, while a steep slope lightly filters the later energy.

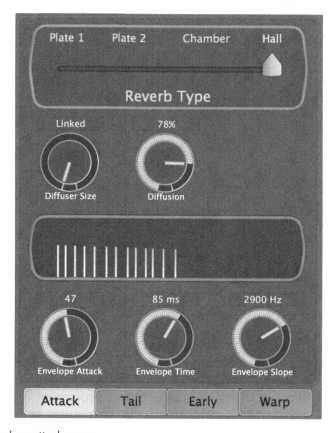

Figure 9.6 *Nimbus*, attack pane.

Source: Used with the permission of Exponential Audio.

The *Early* subpage button opens a pane devoted to adjusting the parameters associated with early reflections (see Figure 9.7). The *Early Attack* dial controls the manner in which the signal enters the plugin (that is, whether the weight of the energy is placed earlier or later). Smaller values produce stronger early reflections, whereas larger values give more weight to the reflections occurring slightly later. The *Early Time* knob adjusts the length of time over which the early reflections are spread, and the *Early Slope* dial lets engineers model air absorption through a low-pass filter that affects the latter part of the early reflections. *Early Pattern* lets the user choose between five distinct groupings of early reflections that range from a fairly even spacing of multiple reflections to a sparse set of three reflections which models vintage hardware processors.

Figure 9.7 *Nimbus*, early reflections pane.
Source: Used with the permission of Exponential Audio.

The *Warp* pane has three rows of dials that determine the dynamic range of the input, add overdrive to the signal, and change the bit depth (word size) of the processing (see Figure 9.8). Four dials control the action of the compressor. *Threshold* sets the level at which the compressor's processing begins, and positive values in *Compression* specify the amount of gain added to the input before compression commences, while negative values control the amount of cut below the threshold. *Attack* controls the length of time between the moment an audio signal exceeds the threshold and the start of the compressor's action, and *Release* sets the length of time it takes for the compressor to cease its processing after the signal falls below the threshold. A gain-reduction meter at the end of the row shows the effect of the compressor.

Three buttons below these dials let users determine the *Knee* function of the compressor, turn a *Limiter* on or off, and through the *Cut* button, set a lower threshold for the compressor. Exponential Audio describes the usefulness of this latter feature:

> Normally, everything below the compressor threshold is boosted by the gain amount (determined by the *Compression* parameter). But often, very low level signals may be only noise—footfalls, breaths, leakage, etc. Any signal below the *Cut* level will not be boosted. For distance mics used in classical music, this should normally be set to a very low value.

(Exponential 2018b: 43)

The four overdrive dials in the second row let users select from among three styles of saturation (*Overdrive Type*—Warm 1, Warm 2, and Crush), set the frequency below which overdrive is applied (*Overdrive Xov*), control the amount of the effect (*Drive*—higher values increase the number of harmonics or overtones), and if desired, reintroduce the input signal above the crossover frequency (*High Passthru*). In the third row, *Word Size* changes the bit depth of the processing from floating-point to 24, 20, 18, 16, 14, or 12 bits, while the *Warp Trim* knob adjusts the overall level of the output from the *Warp* pane.

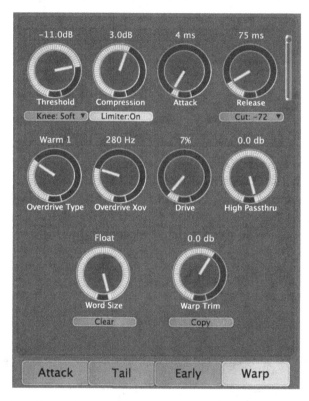

Figure 9.8 *Nimbus*, warp pane.

Source: Used with the permission of Exponential Audio.

In *Verb Session*, Flux simplifies matters somewhat, while maintaining a largely traditional approach to the design of the plugin's GUI (shown in Figure 9.9), that is, a single window divided into four main areas.

The time-structure display in the upper left corner (see Figure 9.10) depicts the main parameters of reverberation. The dotted vertical line on the left represents the dry or direct signal, which is sent straight to the output of the plugin, as it is the first sound listeners hear. The group of vertical bars to its immediate right portrays the initial early reflections, with the spacing between them indicating

Figure 9.9 Flux, *Verb Session* GUI.

Source: Used with the permission of Flux: Sound and Picture Development.

the time locations of the reflections and the height of the bars showing their level (the gap between the direct signal and the initial reflections illustrates the pre-delay). The next set of vertical bars displays the transitional stage between early and late reflections, which Flux calls cluster, and these slightly later, yet still early, reflections are denser in nature and precede reverberation. The solidly colored area depicts the dense wash of late reflections that follows, with the lighter and darker sections showing the decay curves of the frequency bands the plugin generates.

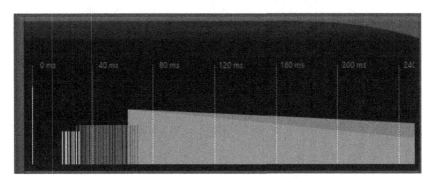

Figure 9.10 *Verb Session*, time-structure pane.

Source: Used with the permission of Flux: Sound and Picture Development.

Below the time-structure pane lies the main controls for the plugin (see Figure 9.11). With the *Room Size* knob, users select the volume of the room to be emulated (in cubic meters), and in the illustration, 15,000 m^3 has been chosen to model the Großer Musikvereinssaal in Vienna (a distinct advantage of Flux's plugin is that users can easily approximate the three main parameters of the concert halls listed in Part 1, Chapter 1—volume, reverberation time, and pre-delay). Once the volume has been determined, *Verb Session* establishes the basic nature of the reverberation in the space (start of the reverb, earliest reflections, cluster transition, and reverb tail). The *Decay Time* knob specifies the length of the reverb tail, which has been set to 2.0 sec, the RT_{60} of the Musikvereinssaal. The plugin suggests a value for the *Pre-Delay* (20.1 ms in this case), which can then be altered to match the hall being emulated (12 ms for the Musikvereinssaal). Longer pre-delays help listeners distinguish between the direct and reflected sound, and these sorts of delays can preserve the intelligibility of the source within the space, especially if large rooms and long decay times have been chosen, for these selections would "otherwise drench the audio material in reverberation" (Flux n.d. b: 7). Flux also provides a way for users to control the gain of both the early reflections (earliest plus cluster) and the reverb tail. The slider under *Early* alters the degree of presence the source has in the room (moving it up or down makes the source seem closer or farther away), while the slider for the *Tail* increases or decreases the level of late reflections in the signal. The *Damping* sliders adjust how rapidly the low- and high-frequency bands within the reverb tail decay. The default setting is 100%, and below this value, the bands die away more quickly but above it they recede more slowly.

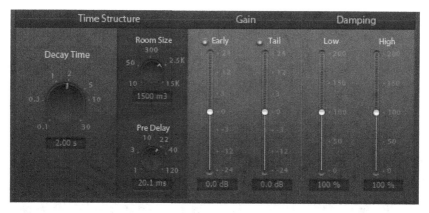

Figure 9.11 *Verb Session*, main controls.
Source: Used with the permission of Flux: Sound and Picture Development.

Engineers alter the tonal quality of the reverb through the filter pane in the upper right corner of the GUI (see Figure 9.12). Three adjustable bands are available (high and low shelves at either end of the frequency spectrum, with a bell in the middle), and users drag the control points up or down to cut or boost the signal. Cut-off points for the shelf filters are set either through the numerical boxes on the right or by dragging the vertical bars back and forth.

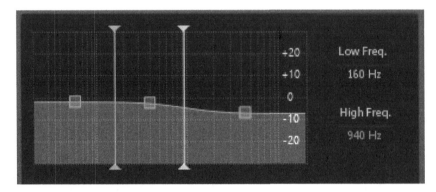

Figure 9.12 *Verb Session*, filter pane.

Source: Used with the permission of Flux: Sound and Picture Development.

Below the filter pane lies the *Input/Output* section of the controls (see Figure 9.13). The *Input* slider determines the level of the signal that enters the plugin, and the *Output* slider lets engineers trim or boost the signal exiting *Verb Session* (trimming can alleviate any clipping the processing may have introduced). The *Dry/Wet* knob designates the mix of untreated (dry) and treated (wet) signal that leaves the plugin (the default setting of 100% means that only the treated signal exits *Verb Session*).

Figure 9.13 *Verb Session*, input/output controls.

Source: Used with the permission of Flux: Sound and Picture Development.

But beyond the more traditional controls offered by Sonnox, Exponential Audio, and Flux lies FabFilter's method of presenting the reflection-simulation algorithms in *Pro-R* (see Figure 9.14). FabFilter focuses on a set of "non-technical" controls (FabFilter n.d. b: 3), that is, knobs for brightness, character, distance, space, decay rate, stereo width, and mix, as well as a graphic section that adjusts the decay time across the frequency spectrum (the upper curve) and equalizes the sound of the final reverb (the lower curve). A real-time spectrum analyzer provides a visual display of the decay times at different frequencies, and a control in the bottom bar varies pre-delay between 0 and 500 milliseconds.

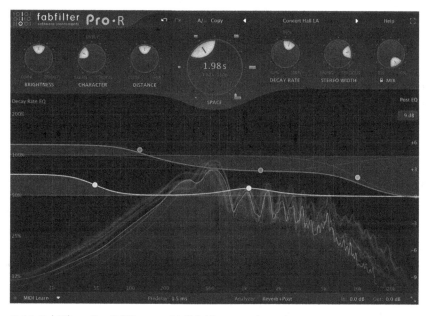

Figure 9.14 FabFilter, *Pro-R* ("Concert Hall LA" preset shown).
Source: Used with the permission of FabFilter Software Instruments.

The presets included in *Pro-R* model a large number of ambient spaces, and with the controls along the top of the plugin, users either design their own ambience or alter the character of FabFilter's rooms. The centrally located *Space* knob changes reverb times seamlessly from those of small studios to those of the largest cathedrals, while the *Decay Rate* control varies whatever setting has been chosen, reducing it to as little as 50% of the original and increasing it by as much as 200%. The *Distance* knob replicates the effect of moving closer to or farther away from the sound source in the selected space (that is, the clarity and presence of early reflections give way to a more diffuse reverb tail), and the *Stereo Width* control shifts the signal from mono (0%) to stereo (50%) and beyond (settings above 50%). The *Brightness* knob governs the balance between the reverb's high and low frequencies, so that engineers can create the darker sonorities found in actual rooms (warmer tonal qualities are a by-product of the absorption of higher frequencies). The *Character* control further alters the style of the reverb from the dense, smooth sound of the clean setting to the livelier color introduced at 50% of the dial's rotation, or as FabFilter describes the "lively" position, "some modulation and more pronounced early reflections together with subtle yet noticeable late echoes" (FabFilter n.d. b: 8). Past this halfway point, the modulations gradually increase in strength, reaching a chorus-like effect at 100%. To change the proportion of wet and dry signal, users turn the *Mix* knob.

FabFilter's approach to EQ also differs considerably from other software developers. Instead of employing a crossover system for modifying the decay rate of low and high frequencies independently of one another (a necessary feature of reflection simulation, as high frequencies decay much faster in rooms than lower ones), *Pro-R* uses a decay-rate EQ. This innovation gives engineers an opportunity to lengthen or shorten the decay time of specific frequencies, selected by clicking on the upper curve to create up to six nodes that can be moved around the screen like a parametric EQ (bell, notch, and shelf curves). Moreover, *Pro-R*'s post EQ feature equalizes the final quality of the ambience through a process that automatically adjusts the gain of the reverb to compensate for the cuts and boosts made with the EQ filters (bell and shelf).

CONVOLUTION

The term convolution refers to the blending or convolving of one signal with another. Reverb plugins operating on this principle combine dry input signals with previously recorded impulse responses to simulate the sound of music performed in ambient spaces. Engineers create impulse responses by recording all the room reflections an initial stimulus generates. A short burst of sound (for example, a starter pistol) or a full-range frequency sweep played through loudspeakers can be used to excite the air molecules in an enclosed space, and after the sound of the stimulus has been removed from the recording (through a process known as deconvolution), the remaining reverberation characteristics, that is, the room's reverb tail or impulse response, can be added to a dry signal.

Since digital audio is primarily a series of discrete measurements of varying amplitude, a computer must make an enormous number of mathematical calculations when it adds an impulse response to 44,100 or more samples every second, while taking into account successive reverb tails as they decay. Hence, convolution reverbs draw heavily on a computer's processing power, and even though this no longer presents problems, engineers often find that unless impulse responses have been created carefully, rather sterile simulations can result.

EastWest has tackled this issue in *Spaces*, a convolution reverb that emulates rooms from studios and smaller concert halls to cathedrals by focusing on the reverberation characteristics produced by specific instruments, instead of the generalized response of a space to an impulse. Consequently, EastWest's recording process has involved positioning loudspeakers to mimic the projection properties of various instruments. For example, when capturing the response of orchestral halls,

a recording was taken in the exact position an instrument would be on stage. A French horn fires its sound backwards against the back wall, therefore, an impulse for a French horn was created by firing the sweep tones from the middle to left rear of the stage, backwards and slightly towards the floor. 1st violin section recordings were taken by firing a series of speakers at an angle toward the ceiling, just like a real section, with a fifth speaker firing towards the floor to emulate the body of the violins.

(EastWest n.d.: website)

This approach has allowed the engineers at EastWest to balance early reflections with room reverberation in a natural way, and Figures 9.15 and 9.16 show the GUI from *Spaces II*. As with many convolution reverbs, the controls are quite simple. Users may select the level of the audio's input, the length of the pre-delay (in milliseconds), the extent of the dry sound that will be included in the audio that exits the plugin, and the output level of the wet signal. They may also adjust the degree of high- and low-frequency roll-off through pass filters and tailor the decay time of the impulse to suit the project.

Figure 9.15 EastWest, *Spaces II*, GUI.
Source: Used with the permission of EastWest Communications.

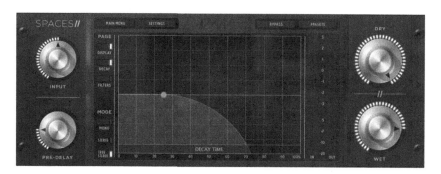

Figure 9.16 *Spaces II*, GUI with decay time pane shown.
Source: Used with the permission of EastWest Communications.

REFERENCES

Case, Alexander U. 2007. *Sound FX: Unlocking the Creative Potential of Recording Studio Effects*. Boston, MA: Focal Press.

EastWest. 2018. *Spaces II*. Los Angeles, CA.

EastWest. n.d. *Website, Spaces II*. Available at: www.soundsonline.com/spaces-II.

Exponential Audio. 2018a. *Nimbus*. Salt Lake City, UT.

Exponential Audio. 2018b. *Nimbus, User Guide*. Salt Lake City, UT.

FabFilter. n.d. a. *Pro-R*. Amsterdam, The Netherlands.

FabFilter. n.d. b. *Pro-R, User Manual*. Amsterdam, The Netherlands.

Flux. n.d. a. *Verb Session*. Orleans, France.

Flux. n.d. b. *Verb Session v^3, User Manual*. Orleans, France.

Sonnox. 2017. *Oxford Reverb*. Finstock, UK.

Sonnox. n.d. *Oxford Reverb, User Guide*. Finstock, UK.

CHAPTER 10

Delivery

FILE TYPES

Containers and Codecs

A container functions as a wrapper that holds digital audio information, along with metadata, whereas a codec, an abbreviation of coder/decoder, is an algorithm designed to compress and decompress digital audio information. In other words, containers specify the organization of the data within them (for example, interleaving video/audio data in chunks for streaming purposes), while codecs provide various ways of storing data so the digital information occupies less space.

Codecs employ one of two compression methods, lossless or lossy. Lossless algorithms compress data to about half the size of the original, yet allow that data to be reconstructed perfectly during decoding. Lossy codecs, however, take a psychoacoustic approach to compression and remove information humans do not hear well. By analyzing the signal and retaining only the most audible components, these codecs create files as much as ten times smaller than the original. Since the discarded information can never be recovered, these algorithms achieve their reductions at the expense of sound quality, especially at lower bit rates.

Common File Types for the Delivery of Audio

Uncompressed

WAV

The Waveform Audio File Format (WAV or WAVE) uses a RIFF container (Resource Interchange File Format) that holds PCM (pulse code modulation) digital audio information. It is one of the two standard uncompressed file types available to recordists (the other is AIFF).

AIFF

Apple's Audio Interchange File Format (AIFF), which is very similar to WAV, also employs a RIFF-type container to house uncompressed PCM digital audio information.

Lossless Compression

FLAC

The Free Lossless Audio Codec (FLAC) was developed by the Xiph.Org Foundation and uses the Ogg container (hence, it is also known as Ogg FLAC). FLAC compresses data without any loss of audio quality, and since Xiph.Org makes it available at no charge, it has become the standard lossless audio codec internationally.

ALAC

The Apple Lossless Audio Codec (ALAC) employs an m4p audio-only container, with the file extension m4a, and compresses audio data with no loss of information. It is supported on iPad, iPhone, iPod, and iTunes, as well as on Mac computers.

Lossy Compression

MP3

The Fraunhofer Institute for Integrated Circuits in Erlangen, Germany, in collaboration with other institutions, has participated in the development of a number of codecs, including the ubiquitous mp3. An mp3 encodes and stores audio so that the data occupies as little as 9%–10% of the original file's storage space (assuming the compression of a 16 bit/44.1 kHz recording).

In 1989, the international standards organization Moving Picture Expert Group (MPEG) was interested in establishing an audio coding standard, and between 1989 and 1991, they encouraged the development of a family of coding techniques, with layer 3 of this family receiving the name mp3 in 1995. Although the audio industry has embraced this codec almost universally, the main problem with the perceptual coding used to create an mp3 has always been the loss of sound quality at lower bit rates. This degradation can include distortion, roughness, and pre-echoes, as well as noise in certain frequency ranges (for further discussion, see Brandenburg n.d.: 6–7).

AAC

The Advanced Audio Codec (AAC), developed by the Fraunhofer Institute in collaboration with other companies as a replacement for the mp3, features new coding tools for improved audio quality at lower bit rates. AAC also employs an m4p audio-only container, and in 2003 Apple launched their iTunes platform based on the Advanced Audio Codec. Confusingly, both AAC and ALAC files use the extension m4a.

VORBIS or OGG VORBIS

The Xiph.Org Foundation created Vorbis as a lossy codec to be used in their Ogg container format. Like mp3, it employs psychoacoustic principles of encoding to reduce the amount of data, but unlike mp3, Vorbis is free and unpatented.

File Size

Table 10.1 shows the file sizes that result when various types of processing (bit depth/sample rate conversion and compression) are applied to a recording of

Table 10.1 File size.

File Type	Details	Size (MB)
WAV (original)	24 bit/96 kHz	108.4
WAV (converted)	16 bit/44.1 kHz	33.0
FLAC	24 bit/96 kHz	61.5
FLAC	16 bit/44.1 kHz	14.2
AAC iTunes+	256 kbps, 44.1 kHz	5.8
mp3	256 kbps, 48 kHz	6.0
mp3	192 kbps, 48 kHz	4.5
mp3	128 kbps, 48 kHz	3.0
Ogg Vorbis	256 kbps, 48 kHz	3.2

Monteverdi's "Sì dolce è'l tormento" from Studio Rhetorica's *Secret Fires of Love* (tenor and Baroque guitar), 3:07 in length. Weiss Engineering's *Saracon* produced the WAV (converted), FLAC, and Ogg Vorbis files, and Sonnox's *Fraunhofer Pro-Codec* generated the AAC iTunes+ and mp3 files.

LOUDNESS AND METERS

Definition of Loudness

In 2013, the Advanced Television Systems Committee (ATSC) in Washington, DC, defined loudness as a perceptual quantity, that is, the magnitude of the physiological effect produced when a sound stimulates the ear (ATSC 2013: 14). These physiological reactions can be quantified (measured) by digital meters employing an algorithm designed to approximate the human perception of level, and the adoption of "loudness" meters in recent years has transformed the broadcast industry.

Standards for Measurement

In response to complaints from consumers about the lack of uniformity in the subjective loudness of audio programs transmitted to audiences, the International Telecommunication Union (ITU) advocated a fundamental change to the way loudness was calculated, and in 2006 the ITU introduced algorithms for objectively measuring both the perceived and the peak level of digital audio signals. These two algorithms not only allowed broadcasters to assess the perception of "audio volume,"[1] instead of simply measuring peak loudness at sampling points, but also provided a fairly accurate estimate of undetected audio peaks occurring between sample points (the true-peak level).

Four years after the ITU made the algorithms available, the European Broadcasting Union (EBU) proposed a metering system that could deal with the various facets of loudness (the latest version of the document appears in EBU 2014).

Beyond measuring the integrated or average loudness of an entire signal, a method that closely relates to the way humans experience overall energy, the EBU recommended target levels for normalizing audio. They also introduced four other descriptors for characterizing signals: maximum true-peak level (so that the audio could comply with the technical limits of digital systems), maximum momentary loudness (based on a sliding window 400 milliseconds in length), maximum short-term loudness (based on a sliding window of 3 seconds), and loudness range (the distribution of loudness, softest to loudest, across a signal). Moreover, the EBU described the effect the new practices could have on broadcasting and summarized the advantages of their recommendations (EBU 2016b: 7):

> It must be emphasised right away that this does NOT mean that the loudness level shall be all the time constant and uniform within a programme, on the contrary! Loudness normalisation shall ensure that the average loudness of the whole programme is the same for all programmes; within a programme the loudness level can of course vary according to artistic and technical needs. With a new (true) peak level and (for most cases) the lower average loudness level, the potential differences between the loud and soft parts of a mix (or the "Loudness Range") can actually be significantly greater than with peak normalisation and peak mixing practices in broadcasting.

A number of countries have since adopted the EBU (or similar) standards for their broadcast industries, and the new practices apply well beyond broadcasting, for as the ITU suggests, "the matter of subjective loudness is also of great importance to the music industry, where dynamics processing is commonly used to maximize the perceived loudness of a recording" (ITU 2015: 7).

Digital meters that conform to the EBU recommendations feature the following elements (the "EBU mode"):

Three time scales:

M (Momentary)—measures maximum loudness in the last 400 ms

S (Short-term)—measures maximum loudness in the last 3 secs

I (Integrated)—loudness over the duration of a complete self-contained audio signal; this measurement uses gating[2]

Plus:

LRA or Loudness Range—indicates the statistical loudness variation, excluding extremes, between the softest part of the signal and the loudest part

Maximum true-peak level—shows the level of any peaks that occur between sample points

These five parameters provide a fairly complete description of an audio signal's loudness, and the method developed to compensate for subjective hearing is called K-weighting (not to be confused with K-System metering, an approach created by mastering engineer Bob Katz). Because people do not perceive low, mid, and high frequencies, even when heard at identical levels, to be of the same loudness, the new meters apply a K-weighted filter to each audio channel to turn objective measurements into subjective impressions (see Figure 10.1). Specifically, the filter approximates human hearing by de-emphasizing low frequencies (to make them less loud) and emphasizing higher frequencies (to make them louder), before it averages the filtered audio.

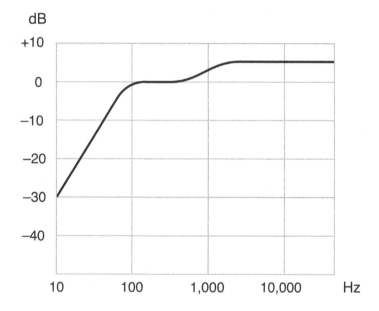

Figure 10.1 K-weighting.

The meters also have relative and absolute scales. The relative scales equate 0.0 dB full scale to −23.0 LUFS, which is the EBU's normalization target (see below for a fuller discussion of target levels and for definitions of the terminology associated with metering). These relative scales, meant for recordists accustomed to older meters, come in two varieties. One has +9.0 dB above zero, so engineers can easily see how far their music peaks above the target level, and the other has +18.0 dB above zero, for more dynamic material. The absolute scale shows recordists the actual readings based on the target of −23.0 LUFS.

Metering Terminology (EBU and ITU)

dBFS (dB Full Scale)—audio level in decibels referenced to digital full scale, that is, referenced to the clipping point ("full scale") in a digital audio system; 0.0 dB represents the maximum level a signal may attain before it incurs clipping.

dBTP (dB True Peak)—maximum inter-sample peak level of an audio signal in decibels referenced to digital full scale, that is, referenced to the clipping point ("full scale") in a digital audio system; 0.0 dB represents the maximum level a signal may attain before it incurs clipping (true-peak meters are discussed below).

LU (Loudness Unit)—a relative unit of loudness referenced to something other than digital full scale; it employs K-weighting and is analogous to dB, for one LU equals one dB; terminology established by the International Telecommunication Union.

LKFS (Loudness, K-weighted, referenced to digital Full Scale)—loudness level on an absolute digital scale; it is analogous to dBFS, for one LKFS unit equals one dB; terminology used by the International Telecommunication Union and the Advanced Television Systems Committee (USA); it is identical to LUFS.

LUFS (Loudness Unit, referenced to digital Full Scale)—loudness level on an absolute digital scale; it is analogous to dBFS, for one LUFS unit equals one dB; LUFS employs K-weighting and is identical to LKFS; terminology used by the European Broadcasting Union, which adopted the LUFS nomenclature over LKFS to "resolve the inconsistencies currently present in ITU-R BS.1770–4 and BS.1771 [4]" (EBU 2016a: 7).

LRA (Loudness Range)—originally developed by TC Electronics, it is the overall range of the material from the softest part of the signal to the loudest part, given in LU; to avoid extreme events from affecting the reading, the top 5% and the lowest 10% of the total loudness range are excluded from the measurement (for example, a single gunshot or a long passage of silence in a movie would result in a loudness range that is far too broad).

Normalization—a method of adjusting loudness so that listening levels are more consistent for audiences.

True-Peak Meters

Since traditional meters measure peak levels only at the sample points (hence, their designation as sample-peak devices), they will miss any peak levels that occur between samples (see Figure 10.2). To address this problem, the algorithm employed in true-peak measurement uses at least 4x oversampling to provide a reasonably accurate estimate of inter-sample peaks. Nonetheless, even with this amount of oversampling, the meters may still underestimate the actual peaks somewhat, so the EBU has recommended that during the production of PCM material engineers should not exceed a maximum true-peak level of –1.0 dB. But if the audio is destined for delivery through a lossy codec using lower bit rates (for example, mp3 with bit rates below 256 kbps), the maximum true-peak level probably should not go beyond –2.0 dB, because codecs employing reduced bit rates need additional headroom to operate, as they often "misread" inter-sample peaks by more than 1.0 dB.[3]

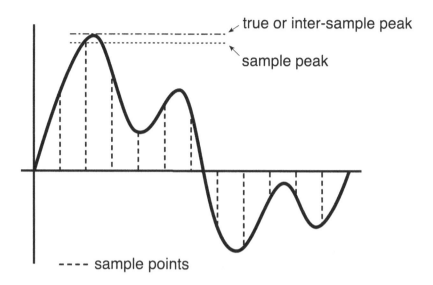

Figure 10.2 True or inter-sample peak vs. sample peak.

Table 10.2 demonstrates the extent to which the true-peak level increases when a representative 24 bit, 96 kHz track (Purcell's "I attempt from love's sickness to fly" from Studio Rhetorica's album *Secret Fires of Love*) is rendered through several codecs or converted for CD distribution. Although the codecs listed here only raise the true-peak levels of this particular example by a maximum of 0.33 dB, and the sample rate converters increase the level by no more than 0.04 dB, the guidelines recommended by the EBU and organizations in other countries (see Table 10.3) provide distortion-free working practices for recordists who wish to leave plenty of headroom as they track.

Table 10.2 Codecs and sample rate conversion: changes in true-peak level (CBR abbreviates "constant bit rate" and VBR denotes "variable bit rate").

Details	TP (dB)
original file—24 bit/96 kHz	−0.75
Codec	
Ogg Vorbis, CBR 128 kbps	−0.42
mp3, CBR 128 kbps	−0.56
Ogg Vorbis, CBR 256 kbps	−0.57

(Continued)

Table 10.2 Continued

Details	TP (dB)
Ogg Vorbis, VBR, quality 10	−0.64
AAC-LC, CBR 96 kbps	−0.66
AAC iTunes+, VBR, quality—highest	−0.72
mp3, CBR 256 kbps	−0.72
SRC/Dither (16 bit/44.1 kHz)	
iZotope/MBIT+	−0.71
Saracon/POWr3	−0.74
Saracon/TPDF	−0.74
WaveLab/MBIT+	−0.74

WaveLab Pro 9.5 generated the Ogg Vorbis files, while Sonnox's *Fraunhofer Pro-Codec* created the AAC and mp3 files. Sample rate conversion was undertaken in iZotope's *Rx 6 Advanced*, Weiss Engineering's *Saracon*, and Steinberg's *WaveLab Pro 9.5*, and the true-peak levels were measured in iZotope's *Rx 6 Advanced*.

Target Levels

Broadcast standards organizations have set target normalization levels for loudness, and as mentioned above, the EBU suggests an integrated loudness level of −23.0 LUFS (±0.5 LU), with a maximum permitted true-peak level of −1.0 dBTP (or −2.0 to −3.0 dBTP for lossy codecs to avoid downstream clipping). The range of practices in various countries are shown in Table 10.3.

The normalization targets for streaming platforms are higher (louder), for in the words of the AES, "some current mobile devices have insufficient gain to allow the common production targets of −23.0 or −24.0 LUFS to be heard at a satisfying loudness, even if the volume control is turned all the way up" (AES 2015: 2), especially in noisy listening environments. While the AES promotes a target of −16.0 LUFS (AES 2015: 2),[4] the levels used in specific systems currently vary by 3.0 dB (see Table 10.4).[5]

Table 10.3 Target values for true-peak and integrated levels.

Region	True Peak	Integrated	Integrated Tolerance (+/-)
Europe	−1.0	−23.0	0.5
USA	−2.0	−24.0	2.0
Australia	−2.0	−24.0	1.0
Japan	−1.0	−24.0	2.0

Table 10.4 Normalization targets of streaming platforms.

Streaming Platform	Normalization Target	Codec
YouTube	−13.0	AAC-LC/Opus
Spotify	−14.0	Ogg Vorbis
Tidal	−14.0	AAC-LC
Apple Music	−16.0	AAC-LC

Loudness in Practice

After the ITU, EBU, and AES published their recommendations, the ultimate goal of loudness normalization has become the harmonization of levels between broadcasts, as well as among the individual components that make up a longer program of material. When applied to music, this generally means achieving uniform loudness levels between the tracks on a disk or in a playlist/stream.

The advantages of this approach for popular music have been summarized by Florian Camerer (2010: 2):

> The fight for "Who is the loudest" disappears, mixes can be more dynamic, there are fewer dynamic compression artefacts, such as "pumping," and thus there is an over-all increase of audio quality! Programme makers who favoured dynamic mixes in the past are now relieved from potential compromises because their programme no longer sounds softer than more-compressed ones. With loudness normalisation, this compromise is gone.

In classical music the ultimate goal, at least according to supporters of the complete package of EBU protocols, is

> to provide a satisfactory listening experience for a diverse mix of genres for the major-ity of listeners. This may result, for example, in a Schubert string quartet having the very same integrated loudness level as a Mahler symphony, . . . [and] while this does not reflect reality, it makes these items fit into a wide array of adjacent programming.
> (EBU 2016b: 30)

And if one sets aside the relatively new concerns over loudness uniformity in broadcasting and streaming, the EBU recommendations correspond to the historic practices encountered on classical records closely enough that the guidelines, especially for integrated levels, suit the genre quite well. Because the dynamics in classical performances routinely extend from *pianissimo* to *fortissimo*, engineers working in the field like to preserve that spectrum on their recordings, even if this makes softer passages hard to hear in noisy environments (such as a car). Indeed, the measurements shown in Table 10.5, which have been taken from a random selection of tracks on CDs issued between 1980 and 2014, exemplify typical procedures. The chart reveals integrated

Table 10.5 Loudness in CDs (measured in iZotope's *RX 6 Advanced, Audio Editor*; TP = true peak, I = integrated, LRA = loudness range).

Composer	Work	Artist/Label & Date	TP	I	LRA
Mahler	Symphony No. 9 (i)	Klaus Tennstedt EMI, 1980	−0.9	−18.2	23.3
Binchois	Deuil angoisseux	Clemencic Consort Musique en Wallonie, 1980	−3.1	−22.9	15.7
Mozart	Requiem, K626 (i)	Christopher Hogwood Decca, 1984	−0.3	−15.5	18.9
Vaughan Williams	Mass in G minor (i)	Stephen Darlington Nimbus, 1987	−5.7	−29.2	21.9
Barber	String Quartet, Adagio	Manhattan String Quartet Newport Classic, 1987	−6.8	−24.2	16.0
Beethoven	Sonata No. 23 (i)	Melvyn Tan EMI, 1988	−0.7	−19.1	18.2
Ward	Come sable night	Hilliard Ensemble EMI, 1988	−6.7	−24.6	15.2
Berlioz	Symphonie fantastique, Op. 14 (i)	Roger Norrington EMI, 1989	−0.3	−23.6	23.5
Mendelssohn	Piano Concerto No. 1 (i)	Christopher Kite Nimbus, 1989	−3.7	−23.2	18.2
Chopin	Ballade, Op. 52	Andrei Gavrilov DG, 1992	−0.2	−21.7	23.0
Vivaldi	L'Atenaide Ferma, Teodosio	Emma Kirkby Hyperion, 1994	−4.0	−24.5	16.5
Vivaldi	Le quattro stagioni L'estate (i)	Il Giardino Armonico Teldec, 1994	−0.2	−22.3	23.6
Rossini	Or che di fiori adorno	Cecilia Bartoli Decca, 1997	−0.5	−23.1	15.8
Handel	La Lucrezia Già superbo	Lorraine Hunt Lieberson Avie, 2004	−2.7	−23.4	22.7
Guédron	Cessez, mortels	Emma Kirkby BIS, 2006	−3.3	−23.3	17.2
Handel	Messiah, Rejoice greatly	John Butt Linn, 2006	−2.2	−22.1	10.8
Handel	Messiah He was despised	John Butt Linn, 2006	−7.2	−27.1	12.4
Bellini	Norma Casta Diva	Cecilia Bartoli Decca, 2007	0.0	−21.2	19.6
Debussy	Preludes, Bk 1, no. 10 La Cathédrale	Hiroko Sasaki Piano Classics, 2010	−0.3	−19.4	26.2
Dowland	Can she excuse my wrongs?	Iestyn Davies Hyperion, 2014	−0.7	−21.4	9.1
Sances	Chiesi un bacio	Bud Roach Musica Omnia, 2014	−0.1	−16.2	11.3
Arena	La clemenza di Tito Come potesti, oh Dio	Simone Kermes Sony, 2014	0.0	−15.7	13.6

levels from –15.5 to –29.2 LUFS (average of –21.9, with sixteen of the twenty-two tracks below –21.2), loudness ranges between 9.1 and 26.2 LU (seventeen of the tracks measure from 15.2 to 23.6), and true-peak values between –7.2 dBTP and full scale (twelve tracks register from 0.0 to –0.9, while ten tracks keep true peaks below the EBU's recommended –1.0, that is, between –2.2 and –7.2). The table includes a few selections that are consistently soft throughout—the movement from Vaughan Williams' *Mass*, Barber's *Adagio*, and Handel's "He was despised" from *Messiah*.

However, the loudness parameters shown in this chart contrast sharply with those found in popular music, and a convenient illustration of the extent to which popular and classical styles diverge in terms of loudness can be found on disks issued by the opera singer Renée Fleming. In 2010, Fleming released *Dark Hope*, a recording of popular music that not only fully embraced the loudness tenets of the genre (an LRA below 8; the LRA in pop recordings commonly lies between 4.0 and 8.0 LU [Katz 2015: 74]) but also differed considerably in this regard from her classical CDs.

Table 10.6 places the loudness characteristics of two tracks from *Dark Hope* beside art songs of a similar length from Fleming's album *Night Songs*. Both "Today" and "With twilight as my guide" combine the narrow loudness range typically found in popular music (7.9 and 7.8 LU respectively) with a relatively high program/average loudness (–9.9 and –9.8 LUFS). But the recordings of "Clair de lune" and "Ruhe, meine Seele!" display the greater dynamic spectrum classical listeners prefer (19.0 and 23.0 LU), as well as program/integrated loudness much lower (–20.4 and –20.0 LUFS) than the average usually encountered in popular genres.

The restricted dimensions of loudness in popular music result, of course, from the amount of dynamic-range reduction (compression) engineers employ, and screen shots of the complex waveforms for "Today" and "Clair de lune" show, at least with regard to loudness range, the distinct sound qualities that contribute to the sonic signatures of the two genres (see Figures 10.3 and 10.4).

Table 10.6 Loudness practices in Renée Fleming's *Dark Hope* and *Night Songs*.

Composer	Work	Label & Date	TP	I	LRA
Balin & Kantner	Today (*Dark Hope*)	Decca Records, 2010 (Avatar Studios, NYC)	0.0	–9.9	7.9
Bixler	With twilight as my guide (*Dark Hope*)	Decca Records, 2010 (Avatar Studios, NYC)	0.0	–9.8	7.8
Fauré	Clair de lune (*Night Songs*)	Decca Records, 2001 (The Hit Factory, NYC)	–2.2	–20.4	19.0
Strauss	Ruhe, meine Seele! (*Night Songs*)	Decca Records, 2001 (The Hit Factory, NYC)	–0.1	–20.0	23.0

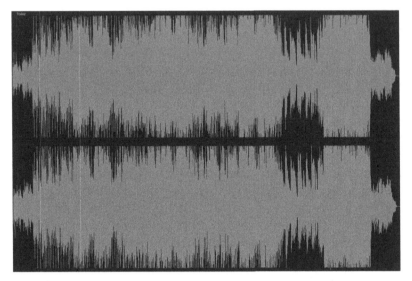

Figure 10.3 Renée Fleming, "Today" (*Dark Hope*).

Source: Screenshot from *Mixbus* used with the permission of Harrison Consoles.

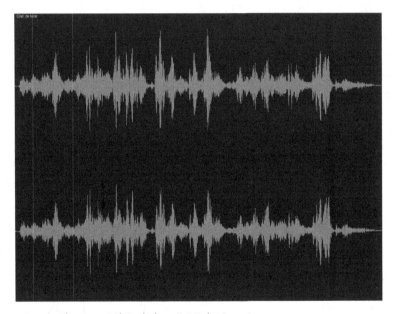

Figure 10.4 Renée Fleming, "Clair de lune" (*Night Songs*).

Source: Screenshot from *Mixbus* used with the permission of Harrison Consoles.

Metering Plugins

A number of companies produce plugins that allow users to measure loudness according to EBU guidelines, with MeterPlugs and Nugen Audio placing several regional standards at the user's disposal so engineers can easily comply with targets set by individual countries. MeterPlugs' *LCAST* supplies presets for EBU R 128 (Europe), ITU BS.1770 (adopted by a number of countries), and ATSC A/85 (USA), as well as for iTunes' *Sound Check*. Nugen Audio's *MasterCheck* has options for television broadcast in Europe, Japan, and USA, plus presets for AES Streaming Practice, Apple Streaming, BBC iPlayer Radio, DAB+ Radio, Pandora, Replay Gain, Spotify, Tidal, and YouTube.

While both companies adhere to EBU recommendations, their meters package the measurements in different ways (see Figures 10.5 and 10.6). MeterPlugs, in addition to EBU loudness categories, provides a graphic "history plot" that displays integrated, short-term, and momentary loudness across an entire track. Nugen Audio, on the other hand, substitutes PLR (defined below) for LRA and includes a procedure for listening to the effect various codecs and streaming platforms would have on the audio.

In Figures 10.5 and 10.6, the same audio track (a "Fantasie" for Renaissance guitar from Studio Rhetorica's *Secret Fires of Love*) has been measured by the two meters in reference to the European television standard (–23.0 LUFS).

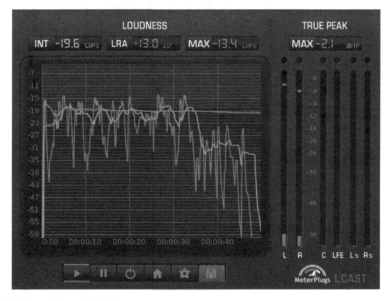

Figure 10.5 MeterPlugs, *LCAST*.

Source: Used with the permission of MeterPlugs.

LOUDNESS: boxes left to right—integrated, range, maximum momentary graphic "history plot"—integrated, short-term, momentary loudness

TRUE PEAK: box—maximum level for the highest channel
below box
—the first two bar meters are L and R stereo channels
—the next four bar meters are surround channels

Figure 10.6 Nugen Audio, *MasterCheck*.

Source: Used with the permission of Nugen Audio.

LEFT BAR METERS:	S = short-term loudness; PLR = peak-to-loudness ratio (the PLR value for the entire track is shown in the box above the second bar meter)
CENTER SECTION:	first box—short-term, real time PLR second box—integrated loudness area below boxes—Encode (codec metering)
RIGHT SECTION:	bar meters—maximum true-peak level in L and R stereo channels below bar meters—individual true-peak meters for codecs

LCAST measures the LRA or loudness range of the signal at 13.0 LU, and *MasterCheck* calculates the PLR or peak-to-loudness ratio to be 17.5 LUFS. PLR is "the ratio between the highest true peak not exceeding 0 dBTP, and the long-term average loudness of the [track] or album in LUFS" (Katz 2015: 223). In other words, since PLR measures the difference between a signal's maximum true-peak level and its integrated loudness, it provides engineers with a useful diagnostic tool: as long as the PLR is lower than the available headroom (Lund & Skovenborg [2014: 3] define headroom as "the ratio between the maximum peak level a signal path or system can handle and its target [average] loudness level"), engineers may perform some sort of static gain correction ("normalization") without incurring clipping or resorting to excessive peak limiting (Lund & Skovenborg [2014: 3]; for example, if the target average loudness level during production is –23.0 dB and the allowable maximum true-peak level is –1.0 dB, then the headroom is 22.0 dB). The two meters measure the integrated loudness of the "Fantasie" at –19.6 and –19.7 LUFS respectively and estimate its highest true peak to be –2.1 dBTP (*MasterCheck* includes numeric readouts for the true peaks of both stereo channels).

Beyond these facets of metering, the *Encode* area of *MasterCheck* allows engineers to hear how their mixes would sound when streamed or delivered through lossy codecs. To take advantage of this feature, the audio would first be measured using the normalization target to which it was mixed (perhaps –23.0 LUFS), then the engineer would select one of the streaming presets and click *Offset to Match* and *Monitor* to hear in real time the effect the gain change would have on the track.

In Figures 10.7 to 10.9, the "Fantasie" measured above has been offset to conform to the "AES Streaming Practice" (I use their lower figure of –18.0 LUFS; see Note 4 above), "Apple Streaming" (–16.0 LUFS), and "Spotify" (–14.0 LUFS). *MasterCheck* shows the amount of gain change it has applied to bring the audio from the original integrated value of –19.6 LUFS to the new target level, and in the first example (Figure 10.7) the small increase keeps the maximum true-peak level below full scale. But in the other two illustrations (Figures 10.8 and 10.9), the larger boosts drive the audio beyond 0.0 dBFS and clipping occurs. To prevent this distortion, engineers either would need to employ a limiter or they would have to remix the track to bring the maximum true-peak level down by the appropriate amount.

Figure 10.7 *MasterCheck*, AES streaming practice (offset gain of +1.7 LU; both channels remain below 0.0 dB).

Source: Used with the permission of Nugen Audio.

Figure 10.8 *MasterCheck*, Apple streaming (offset gain of +3.6 LU; left channel clips).
Source: Used with the permission of Nugen Audio.

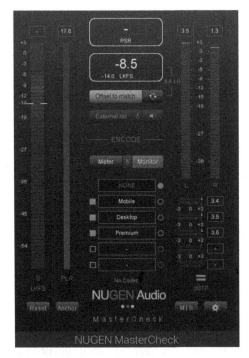

Figure 10.9 *MasterCheck*, Spotify (offset gain of +5.6 LU; both channels clip).
Source: Used with the permission of Nugen Audio.

NOTES

1. This term, borrowed from AES 2015: 1, refers to the subjective combination of level, frequency, content, and duration.
2. The Audio Engineering Society (AES) defines integrated loudness as "a measurement of the total amount of audio energy between two points in time divided by the duration of the measurement. The measurement is frequency-weighted to approximate the sensitivity of the ear to different frequencies and is level-weighted to emphasize the parts of the program contributing most to the sensation of loudness" (AES 2015: 2).
3. Nugen Audio recommends −3.0 dBTP (see Nugen Audio 2016: 3). According to research carried out by Søren H. Nielsen and Thomas Lund in 2003 (p. 10, Table 4), maximum peak values during signal conversion of tracks near full scale could range between +0.3 and +5.3 dBFS (based on a selection of codecs available and tested at the time).
4. AES (2015: 2) further states that "a lower Target Loudness [−18.0 instead of −16.0 LUFS] helps improve sound quality by permitting the programs [that is, complete audio signals] to have a higher peak-to-loudness ratio without excessive peak limiting."
5. These are the target levels used in Nugen Audio's *MasterCheck*. The information on codecs has been taken from Nugen Audio's web article "Mixing & Mastering."

REFERENCES

Advanced Television Systems Committee (ATSC). 2013. *A/85:2013: ATSC Recommended Practice: Techniques for Establishing and Maintaining Audio Loudness for Digital Television*. Washington, DC: Advanced Television Systems Committee.

Audio Engineering Society. 2015. *Technical Document AES TD1004.1.15–10: Recommendation for Loudness of Audio Streaming and Network File Playback: Version 1.0*. Available at: www.aes.org/technical/documents/AESTD1004_1_15_10.pdf.

Brandenburg, Karlheinz. n.d. "MP3 and AAC Explained." Available at: www.iis.fraunhofer.de/content/dam/iis/de/doc/ame/conference/AES-17-Conference_mp3-and-AAC-explained_AES17.pdf.

Camerer, Florian. 2010. "On the Way to Loudness Nirvana—Audio Levelling with EBU R 128." *EBU Technical Review*. Available at: https://tech.ebu.ch/docs/techreview/trev_2010-Q3_loudness_Camerer.pdf.

European Broadcasting Union (EBU). 2014. *EBU R 128: Loudness Normalisation and Permitted Maximum Level of Audio Signals*. Geneva, Switzerland: European Broadcasting Union.

European Broadcasting Union (EBU). 2016a. *Tech 3341: Loudness Metering: "EBU Mode" Metering to Supplement EBU R 128 Loudness Normalization: Version 3.0*. Geneva, Switzerland: European Broadcasting Union.

European Broadcasting Union (EBU). 2016b. *Tech 3343: Guidelines for Production of Programmes in Accordance with EBU R 128: Version 3.0*. Geneva, Switzerland: European Broadcasting Union.

International Telecommunication Union (ITU). 2015. *ITU-R BS.1770–4: Algorithms to Measure Audio Programme Loudness and True-Peak Audio Level*. Geneva, Switzerland: International Telecommunication Union.

Katz, Bob. 2015. *Mastering Audio, the Art and the Science*. 3rd ed. Burlington, MA: Focal Press.

Lund, Thomas and Esben Skovenborg. 2014. "Loudness vs. Speech Normalization in Film and Drama for Broadcast." Paper presented at the SMPTE Annual Technical Conference & Exhibition. Available at: www.aes.org/technical/documentDownloads.cfm?docID=521.

MeterPlugs. n.d. *LCAST*. Vancouver, Canada.

Nielson, Søren H. and Thomas Lund. 2003. "Overload in Signal Conversion." Paper presented at the AES Convention (Copenhagen), 23–25 May. Available at: http://cdn-downloads. tcelectronic.com/media/1018176/nielsen_lund_2003_overload.pdf.

Nugen Audio. 2016. *MasterCheck, Operation Manual*. Leeds, UK.

Nugen Audio. n.d. "Mixing & Mastering for Today's Music Streaming Platforms." Available at: www.nugenaudio.com/mix-and-master-for-streaming-services.php.

PART 4

Common Recording Strategies

This final section considers a number of common strategies for recording acoustic sound sources through microphones, and it covers solo piano, soloists with piano accompaniment, and small ensembles, as well as studio techniques that can enhance or replace the capture of performances in ambient spaces. The descriptions of miking contained in the following chapters should, however, be regarded as guidelines, suggestions for microphone placement to be modified according to the desired soundscape for a project. Since recording has always been a highly subjective art, no single procedure should be regarded as the correct (or only) way to represent the character of either an individual sound source or a group of sources. Indeed, the decisions engineers and producers make in preparation for tracking determine, to a large extent, the nature of a recording's sonic surface, and if the main goal is to satisfy a wide cross-section of the public, recordists must use their critical listening skills to craft a product that will suit people who have differing notions of what constitutes "good" sound.

In fact, the cumulative experience of everyone involved in a project helps them achieve a compelling balance between the spatial environment of the recording, the clarity and transparency of details, the lifelike portrayal of timbre, the faithful representation of constantly changing dynamics, and the realistic rendering of the sound sources in the stereo playback field. Finely honed analytical skills, then, which have been built, in part, on the concepts treated in the preceding sections of this book, assist producers, engineers, and performers in communicating their ideal conception of the music (in its recorded form) to audiences.

CHAPTER 11

Solo Piano

A sound source as large as a full-size grand piano presents engineers with numerous alternatives for turning the sonic character of the instrument into a stereo recording, and to capture a balanced sound from a source that has diverse possibilities for microphone positioning, many believe that a homogeneous tonal quality can be found only when the soundwaves have propagated beyond the near field and coalesced into a unified image (for large instruments, this distance is somewhat greater than for smaller sound sources). Other engineers, however, prefer to capture the instrument from a closer perspective and adjust their miking techniques accordingly. But in both approaches, recordists generally strive for a microphone setup that situates the piano in the center of the stereo image, more or less half left and half right, and engineers usually locate the higher frequencies (the player's right hand) on the left side of the playback image and the lower frequencies on the right, as this portrays the instrument from the audience's viewpoint instead of the player's.

The sheer size of the grand piano complicates matters, and recordists often walk around the instrument, listening carefully, to determine what the mics will "hear" at various places in the near field. When our ears are in the same location as a pair of microphone capsules, we quickly realize that the piano exhibits quite different tonal qualities along the path from the keyboard to the tail. The transfer of energy from the instrument's vibrating strings to its wooden diaphragm generates soundwaves that radiate from both sides of the soundboard, and the raised lid of the instrument directs the waves traveling up from the soundboard into the room. Since longer wavelengths readily bend around objects, the lid reflects mid and high frequencies better than lower ones. Hence, for the upper range of the frequency spectrum, the lid acts as a barrier, which causes the tonal quality of the instrument to be considerably muted above and behind the lid.

Apart from this phenomenon, the largest amount of high-frequency information stems from the vicinity of the hammers, where the shortest strings lie, but in the middle of the piano, where the longer strings cross over one another, more mid-frequencies can be found. Towards the tail, a greater proportion of lower frequencies radiate from the instrument, so a warmer sound can be heard there. And if engineers want listeners to experience a bright, yet reasonably balanced signal, they often explore microphone positions within the strong lobe of higher frequency energy, roughly 1.0 to 4.0 kHz, that spreads across a narrow arc in front of the piano (about 10 degrees in width), generally above the halfway point between the lip of the case and the edge of the fully open lid (for further discussion of these phenomena, see Meyer 2009: 163–7).

Recordists also pay close attention to the location of the instrument in the recording space, for reflections from the walls, floor, and ceiling can compromise sound quality. A position adjacent to a wall or too close to the corner of a room may create unwelcome early reflections or boost lower frequencies, and even if the piano is placed a few meters (yards) from a wall, higher frequencies bouncing off a hard surface could cause comb filtering if the reflected waves arrive at a pair of microphones out of phase with sound traveling directly from the source. The floor, however, being closer to the mics than the walls or ceiling, usually generates the greatest number of early reflections, and engineers frequently place a rug under the instrument to reduce or eliminate phase anomalies coming from below.

But beyond considering questions of sound propagation, recordists must decide which transducers are capable of picking up the entire frequency range of the fundamentals emanating from an 88-key piano (27.5 to 4,186 Hz, plus harmonics above 10 kHz for the highest notes of the instrument), as well as which mics have the best off-axis response (most soundwaves from such a large sound source will not arrive at the microphones' diaphragms on-axis). Indeed, to achieve a uniform tonal quality across the piano's frequency spectrum, engineers regularly choose omnidirectional microphones. Cardioids can, of course, also produce a balanced sound, but they need to be located farther from the instrument. In situations that call for close placement, however, the proximity effect inherent in cardioids can make them a less attractive option.

RECORDING IN STEREO

Since the goal of most classical recordists centers on capturing the grand piano in an ambient setting, engineers usually consider the options for stereo miking available to them before experimenting with the position of the chosen array. A decision on the most advantageous location for the mics takes at least three factors into account: tonal balance between the instrument's registers, an acceptable ratio of direct to ambient sound, and a realistic stereo image. Distant positions, of course, capture more of the reverberant characteristics of the room and less of the piano's details, while close placements often highlight one area of the instrument over others, which can lead to an inconsistent presentation of the piano's overall sound. Three well-established techniques remain prized by recordists, an ORTF near-coincident array, A-B spaced microphones, and a mid-side coincident pair.

ORTF

In large rooms that have a pleasing ambience, near-coincident arrays can yield a sound containing enough higher frequency reflections from the lid to produce a satisfying tonal quality. An ORTF pair of mics may be located anywhere from 70 centimeters (27.5 inches) to 3 or 4 meters (10 to 13 feet) in front of the piano at a height of about 1.5 to 3 meters (5 to 10 feet), depending on how close the mics are to the instrument. Recordists direct the array at the piano along a sight line below the fully opened lid, and they often substitute omnidirectional microphones for cardioids. Moreover, many engineers find that a position slightly to the right of the instrument's curve produces a well-balanced frequency spectrum, especially in closer placements.

A-B Spaced Pair

Another common technique involves an A-B spaced pair of omnidirectional mics, with the distance between the mics typically ranging from 20 to 61 centimeters (8–24 inches), the exact spacing being determined by the width of the stereo image one wishes to create, the location of the pair along the piano's perimeter, and the distance from the instrument. Common locations are:

1. perpendicular to the piano, more or less centered on the hammers to capture a great deal of high-frequency information (the mics are often spaced 30–40 centimeters or 12–16 inches apart)
2. at or near the midpoint between the hammers and the tail, pointing into the curve of the instrument, which can yield a pleasant blend of high-, mid-, and low-frequency information; the mics are commonly located an equal distance from a single point at the back of the lid (such as the hinge closest to the player)
3. at the tail of the piano, on either side of the middle line of the frame's casting, roughly 20–30 centimeters (8–12 inches) apart; this provides a warmer sound.

Generally, moving the mics from the tail around to the front of the piano increases the amount of mid- and high-frequency information, and recordists usually experiment to find the location on the perimeter that best suits the engineer/producer/artist's conception for the project. In all three scenarios, the mics may be placed closer to or farther from the instrument, depending on the balance between direct and reverberant sound deemed appropriate. Recordists regularly determine the height, which varies according to the distance the array is from the piano, by sighting along an imaginary straight-line trajectory below the edge of the lid, with the mics angled down about 15 degrees. For the tail position, engineers often place the array 90 to 120 centimeters (3–4 feet) from the instrument.

Mid-Side Coincident Pair

The mid-side technique allows engineers to adjust the width of the stereo image to taste after tracking has been completed, and many recordists find this approach works well when the mics are located somewhere near the curve of the piano fairly close to the instrument.

UNFAVORABLE ROOM ACOUSTICS

Engineers often have to record in spaces with less than ideal reverberation characteristics, and because of this, they sometimes choose to put the microphones inside the piano to minimize or eliminate the negative effects of the room. As noted above, closely placed mics can isolate a narrow section of the instrument, and to lessen the spotlighting effect, recordists broaden the coverage of the microphones by positioning them at least 20 to 25 centimeters (8 to 10 inches) above the strings. Some engineers even place the mics as high as 61 centimeters (24 inches), for this increased distance captures the broadest spectrum of fundamentals and harmonics radiating from the soundboard.

Spaced Pair

Engineers regularly locate one omnidirectional mic over the treble strings (an octave above middle "C" or at the midpoint of the group of shorter strings) and the other above the midrange and bass strings to produce as even a representation of lower frequencies as possible (many feel a good position lies in the area where the strings cross one another). Experimentation with differing heights and mic angles will determine the ideal placements for the envisioned sound ideal.

Coincident Pair

When X-Y cardioids are angled slightly farther apart, perhaps 110 to 115 degrees, and positioned somewhere between 45 and 61 centimeters (18 to 24 inches) above the hammers, the problems associated with accentuating the spot directly beneath the mics can be alleviated, for the greater angle allows the diaphragms to pick up more of the instrument's upper and lower registers. Similarly, a pair of bi-directional mics, in Blumlein configuration over the hammers, can provide an even coverage of the treble to bass registers. Locating both these coincident pairs directly over the hammers produces a brighter, more percussive sound, while placing them somewhat farther along the strings provides greater warmth. Recordists invariably supplement these arrays with a single mic positioned farther down the soundboard over the bass strings.

ORTF Near-Coincident Pair

A pair of cardioid mics in a quasi-ORTF array placed over the strings at an appropriate height and with the mic's angles adjusted to provide an even coverage of upper, mid, and low frequencies can also yield an acceptable sound. Many recordists find that a location somewhere in the middle of the piano's soundboard works well.

REFERENCE

Meyer, Jürgen. 2009. *Acoustics and the Performance of Music, Manual for Acousticians, Audio Engineers, Musicians, Architects and Musical Instrument Makers*. 5th ed. Trans. Uwe Hansen. New York, NY: Springer.

CHAPTER 12

Soloists With Piano Accompaniment

Simultaneously recording a grand piano and a smaller, quieter source (such as voice or violin) can present problems for sound engineers in concert and recital halls. Large, loud instruments produce far more energy than smaller, softer sources, and because of this, engineers try to find locations for microphones that compensate for the dominance of pianos in ambient spaces. Over the years, various strategies for capturing a balanced, homogeneous image of soloists accompanied by pianists have emerged, and these approaches range from a single stereo pair of microphones situated a distance from performers seated or standing in normal concert position to a main stereo pair, augmented by spot mics, with the performers located in places that, as some would argue, better suit recording. This chapter discusses several common techniques.

VOICE

Many singers prefer their voices to "develop" in the room before microphones capture the sound, and a pair of omnidirectional mics in an A-B array, set in an ideal spot to achieve a balance between the voice and the piano, can produce a homogeneous image when the performers are in typical concert positions. Evaluation of locations around the critical distance will find the best proportion of direct to reverberant sound for the project, but if the point chosen in the room does not generate enough presence in the recording, especially for the singer, a spot mic or mics can increase the sense of intimacy.

Other recordists, however, would rather strike the ideal balance between singer and accompanist after tracking has been completed, and these engineers not only employ pairs of mics for both parties but also position the artists to reduce the amount of bleed between the microphone locations. If the stage is wide enough (that is, if early reflections from adjacent walls do not cause the sound quality to deteriorate through comb filtering), a common strategy involves pointing the piano towards the rear of the stage and placing the singer on the opposite side of the raised lid, facing forward into the hall. In this configuration, the lid acts as a barrier, and although the piano's lower frequencies will bend around the lid and reach the singer's mics, higher frequencies will largely be stopped by it.

Engineers regularly employ stereo pairs of omnis in this method, as they pick up the direct sound from the sources, as well as the room's reverberation. Recordists usually set the vocalist's mics behind the music stand, a little more than arm's length from the performer (the stand should be angled appropriately so that reflections from it do not enter the microphones), and if singers keep their heads midway between the mics, any plosive blasts of air will not strike the diaphragms directly. The piano's microphones may be placed in whatever location achieves a pleasing ratio of direct to reverberant sound, as well as the desired blend of voice and piano in the playback image.

Another approach involves a spot mic or mics and a main stereo pair. In this arrangement, the singer faces the center of the piano a distance from it, and the engineer positions an A-B pair of omnidirectional microphones between the performers, with the mics pointing down and spaced apart approximately 60 to 70 centimeters (24 to 28 inches). Recordists find the precise location for the array by moving it back and forth and up and down until the desired symmetry between vocalist and accompanist has been achieved. Many believe the most natural balance occurs when the mics are closer to the singer, for such a placement produces a playback image in which listeners perceive the piano to be slightly farther away than the voice. But if the engineer feels that the vocal quality lacks sufficient presence, a single or pair of cardioid spot mics placed at head height 60 to 100 centimeters (2 to 3 feet) from the singer will correct the deficiency (at these distances, sibilance and proximity effect are rarely problematic).

VIOLIN

Similar complicating factors affect other small sound sources, and recordists have adopted virtually identical miking solutions for violin and piano. But if the violinist stands in the usual concert position adjacent to the keyboard (on the pianist's right), engineers may find it difficult to capture a balanced perspective, for a single pair of stereo mics centered between the violinist and the tail of the piano will inevitably produce a stereo playback image that has the violin occupying much of the left side of the sound stage and the piano filling the right. Locating the microphones elsewhere can alleviate this problem, but to find a better spot engineers need to consider how the violin emits sound. Because higher frequency energy above about 1.0 kHz radiates from the front of the violin and not the back (see the discussion and diagrams for radiation patterns in Meyer 2009: 273–4), setting the pair of mics on either end of an arc from the piano's keyboard to tail probably will not result in a satisfactory recording. For example, a position behind the violinist, with the mics pointing at an angle toward the two players, would not capture direct sound from the violin, as the instrument's soundboard faces the hall and not the microphones. Similarly, a location near the tail will have more piano than violin in the two channels. Recordists might have more success, however, if the violinist moves to the curve of the piano, with the soundboard of the violin pointing into the hall, for this would allow an array situated in front of the instruments to receive direct sound from both sources. To find not only an appropriately balanced stereo image but also the most suitable distance and height for the pair, engineers would once again begin by experimenting in the vicinity of the critical distance.

But if recordists decide to introduce a second pair of mics into the normal concert arrangement, separate stereo arrays can be used for the violinist and pianist.

Experimentation with both the height above and the distance in front of each instrument will determine the best blend of intimacy/presence and room sound.

Other recordists abandon the typical concert arrangement altogether, opting to face the violin towards the piano, and in this scenario, they set the main stereo pair between the instruments in a location that favors the smaller sound source. The microphones are often separated by 60 to 70 centimeters (24 to 28 inches) and placed pointing down from a height of 2 to 3 meters (7 to 9 feet). Engineers move the array from left to right to find the most natural stereo image for the piano, before centering the violinist between the two microphones at a distance that yields the desired ratio of violin to piano. But if this arrangement lacks the sort of presence or intimacy the recordists seek, cardioid spot mics may be added to both instruments.

CELLO

In concerts, cellists usually sit in the curve of the piano, and as the cello faces the audience, an A-B pair of omnidirectional microphones centered in front of the performers at an appropriate distance and height can capture a natural image of the two instruments. If engineers would like a more intimate cello sound, supplementary cardioids can be added, and even though these mics would also point toward the piano, they will be down low enough to avoid some of the reflections from the lid. But if recordists feel the spot mics are capturing too much of the accompaniment in this position, then the cellist could be seated facing the piano, so that the null at the rear of the cardioid spots can help reduce the leakage from the larger instrument. In fact, all the principles of microphone placement associated with positioning soloists in this way, along with the other techniques described above (such as A-B arrays situated between the performers), apply not only to the cello, violin, and voice but also to the instruments discussed below. Hence, to avoid repetition, the succeeding sections will focus on the soloists.

CLARINET, OBOE, FLUTE

Clarinets and oboes radiate sound in similar ways, and spot mics should be placed to capture the full range of the instruments. Most of the frequency information (below 3.0–4.0 kHz) radiates through the unstopped finger holes on the bodies, with the higher frequencies being emitted from the bell (from about 5.0 kHz in clarinets and 8.0 kHz in oboes; Meyer 2009: 297–301). Engineers often locate a spot microphone in front of the instrument with the mic pointing at the lower hand of the player about 61 centimeters (2 feet) away. If the recordist has chosen a cardioid spot, this distance keeps the instrument just outside the point where proximity effect can become an issue. Moving the mic up/down and closer/farther from the source will adjust the sound quality to suit the requirements of the project.

Flutes present engineers with a number of options for capturing the frequency spectrum of the instrument, and since the greatest amount of sound radiates from the embouchure hole, many recordists choose to position a microphone there. But because flutists blow air across the mouth piece to activate the column of air within the instrument, this location not only can be prone to wind blasts but also can sound overly breathy. Nonetheless, a mic in the vicinity of the embouchure hole remains a preferred technique, and to deal with the shortcomings, engineers often position the microphone several centimeters (inches) higher than the flute, employ a pop screen to shelter the diaphragm, and keep the player at least 61 centimeters (2 feet) away.

A second microphone further along the body of the instrument can broaden the sound quality, and to avoid detrimental phase cancellation between the mics, particularly comb filtering, recordists should observe the three-to-one principle discussed at the end of Chapter 4 in Part 2. Many point the microphone towards the midpoint of the body to capture a balanced frequency spectrum from the finger holes.

TRUMPET, TROMBONE, FRENCH HORN

Brass instruments radiate sound from their bells, and higher frequencies tend to propagate directly forward, while lower frequencies spread out over a wider angle (see the discussions and diagrams of radiation patterns for trumpets, trombones, and French horns in Meyer 2009: 305–20). Engineers, then, place microphones according to the quality of the sound they wish to capture. To achieve a balanced frequency spectrum for trumpets and trombones, many recordists prefer to allow the range of emitted frequencies to coalesce before the soundwaves strike the diaphragm, which means situating a spot mic at least a meter (yard) or more in front of the instrument. They also frequently place the microphone off-axis, either above or below the direct line from the bell, so as to avoid the beaming effect of the higher frequencies. An on-axis placement will produce a much brighter (some would say piercing) tonal quality but will sacrifice the warmth created by the more balanced radiation captured off-axis.

Microphone choices range from condensers to ribbons, and care should be taken to select a condenser that does not have too much of a boost in the upper frequencies, for these mics will exaggerate the upper overtones. In fact, many recordists opt for either large diaphragm condensers or ribbons, because these mics have a somewhat more subdued sound (the frequency response of ribbons generally rolls off above about 15 or 16 kHz). Hence, both large diaphragm condensers and ribbons tend to tame some of the more piercing aspects of the tonal qualities emanating from trumpets and trombones.

The bell of the French horn, on the other hand, normally faces away from the audience, so a microphone position in front of the player will present the instrument from the listener's perspective. But if the engineer desires more detail and clarity in the mix, a second microphone behind the player will provide the missing attributes. Other recordists like to place a mic above the player to capture a blend of the fore and aft positions, that is, a mix of clarity and roundness.

REFERENCE

Meyer, Jürgen. 2009. *Acoustics and the Performance of Music, Manual for Acousticians, Audio Engineers, Musicians, Architects and Musical Instrument Makers*. 5th ed. Trans. Uwe Hansen. New York, NY: Springer.

CHAPTER 13

Small Ensembles

PIANO TRIO

When a violinist and cellist join a pianist to record a piano trio, many engineers use a single A-B array of omnidirectional microphones in front of the performers to create a stereo playback image. However, in the typical concert seating arrangement, the mics will be closer to the violin and cello than the piano, and since this discrepancy can make the accompaniment sound slightly farther away, recordists often achieve a more balanced depth perspective by positioning the mics closer to, yet high enough above, the string players to lessen the distance between the strings and the piano. But if engineers find that this approach still does not produce enough accompaniment in the mix, a judicious use of spot mics quite close to the piano will increase the presence of the instrument.

Other recordists choose to place the string players in a location they deem more suitable for recording, and in this scenario, as in projects involving a soloist with piano accompaniment, the violinist and cellist face the piano. Supplemental spot mics can be used to provide clarity and presence for the three instruments, while a main A-B pair of omnis situated in an ideal position between the piano and the strings gives the overall sonic impression of the ensemble.

STRING QUARTET

The members of string quartets have great expertise in creating a natural balance within their ensembles during concerts, and engineers usually employ miking techniques that preserve the tonal symmetry the players have achieved. In a typical seating arrangement, the two violins sit on the left side of the group, with the first violinist closer to the audience, and the cello and viola sit on the right, with either the violist or the cellist nearer the audience. If recordists opt to situate an A-B pair of omnis somewhere in front of the performers to replicate the internal balance of the ensemble on playback, these sorts of positions can lead to an uneven front-to-back image, as the players farther from the mics may sound a bit too distant.

For more uniform coverage, engineers often place the mic stand in the center of the quartet, with the two omnis, spaced 50 to 75 centimeters (20 to 30 inches) apart, pointing at the floor from a height of 3 to 4 meters (10 to 13 feet). By adjusting the distance between the microphones carefully, an equal amount of signal for each player can be captured. But if reflections from the floor have a negative impact on the tonal quality, recordists can raise the mics or place a carpet beneath the performers. Occasionally, engineers set up spot mics for each instrumentalist.

Alternatively, a number of recordists position a Decca tree above the quartet. In this technique, the two outer mics will favor the players below them in the front row, while the middle microphone can be located to pick up equal amounts of the instruments in the second row. During mixing, the Decca tree lets engineers create a sense of either depth perspective or uniformity, for the performers at the front of the ensemble can be made to appear more or less prominent than the players behind them.

CHAMBER CHOIR

Depth perspective is an important consideration when miking choirs, as well, for although vocal ensembles comprised of several singers per part generally project a well-blended image, engineers must consider the discrepancy in distance between the microphones and the front and back rows of the choir. To compensate for two or three tiers of singers, recordists often place an ORTF or A-B array fairly high, at least a meter (3 feet) above the people in the front row, with the mics pointing down at an angle calculated to create an appropriate front-to-back balance. However, if some unevenness remains, risers for the singers at the rear of the choir will bring those individuals closer to the microphones.

With regard to the capture of room sound, placing the chosen array within a meter or two (3 to 6 feet) of the choir will result in a drier quality, as less of the room's reverberation will be "heard" by the microphones, but if the goal is to produce a luxuriantly ambient recording, then either the main pair should be positioned farther back in the hall or a second array should be employed to pick up the reverberant nature of the space. However, when recordists opt for a single pair of mics, a distant location may weaken the blended texture projected by the ensemble, as bass frequencies, because of the lengths of their soundwaves, do not reverberate in concert halls and churches as strongly as higher frequencies do.

CHAPTER 14

Sessions

SOLO PIANO

Spaced Pair at the Tail of the Piano

This technique, much prized in the UK, has been used by Simon Eadon in his recordings of the pianist Marc André Hamelin for Hyperion. A description of the 20 December 2006 session for Hamelin's recording of Charles-Valentin Alkan's *Troisième recueil de chants*, Op. 65 at Henry Wood Hall in London, can be found online, along with two photos of the microphone setup.[1] A spaced pair of Schoeps mics (MK 2S omnidirectional capsules), 23 centimeters (9 inches) apart, were pointed at the tail of the piano and angled slightly down from a height of 168 centimeters (66 inches) a meter or so (several feet) from the instrument. In addition, the lid was raised somewhat higher than usual by a 3-foot (0.9-meter) rod (see Figure 14.1).

Figure 14.1 Microphone setup at the tail of a piano.
Source: Used with the permission of Simon Eadon.

Henry Wood Hall is a large room approximately 10 meters (33 feet) high, 20 meters (65.6 feet) wide, and 33 meters (108 feet) long, and Eadon captured its spacious acoustic in a way that retains clarity, while also providing a great deal of warmth.

Combination of Techniques

Some recordists prefer to blend the distant and close perspectives in their projects, and a readily accessible example of this approach can be seen and heard in a video made at the University of Surrey's Institute of Sound Recording (Keller 2013). The video describes the techniques used to capture a performance of Ravel's *Sonatine*, No. 2 in Studio 1, which has a 1.1 to 1.5 second reverberation time across 250 square meters (2,690 square feet) of floor space. The microphone setup consisted of two AKG C414s in Blumlein configuration over the hammers, two AKG C414s in a mid-side array facing the lid of the piano (to the right of the curve), a ribbon mic below the tail end of the soundboard to provide additional warmth, and for room ambience, a pair of Schoeps mics (MK 2H omnidirectional capsules) spaced about 30 centimeters (12 inches) apart and located a distance from the piano. Beyond this setup, the engineer augmented the room's ambience with a Lexicon 480L external reverb unit.

SOLO CELLO

When Ron Searles (Red Maple Sound) recorded Winona Zelenka playing the Bach cello suites on an instrument made by Joseph Guarnerius in 1707, he chose Royer ribbon microphones to capture the sound of the cello in what he felt was the most natural way.[2] The recording took place at Toronto's Pong Studio, and because Searles liked the smooth top end of ribbons, a feature that helps him preserve the character of bowed strings without exaggerating the upper frequencies in a harsh way, he tracked Zelenka through matched sets of Royers arranged in both a modified Decca tree (R-122) and an A-B stereo pair (R-122V; the A-B pair were not combined with the Decca tree but were for a separate recording on a Studer A80 two-track analog tape machine). All five mics pointed down at the instrument, with the center microphone of the tree formation located 3 to 4 feet (1 to 1.25 meters) from the cello, while the two outer mics were positioned another couple of feet (61 centimeters) away. He placed the A-B pair between the microphones of the tree's broad triangular arrangement (see Figure 14.2).

Since the amount of signal contributed by the center and outer mics of a Decca tree can be adjusted, Searles was able to tailor the width and focus of the stereo image to suit his project, and beyond this, the figure-8 polar pattern of ribbons (equally strong front and rear lobes, with nulls at 90 degrees) helped him reduce phase problems caused by the short delays between the direct sound from the cello and reflections arriving from the side walls. Moreover, the posterior lobes of the Royers picked up the denser reverberation occurring farther back in the room (longer time lags that do not affect the sound quality negatively).

His techniques have produced a recording that makes listeners feel as though they are hearing the intimate and warm sound of the cello in the same space as the performer.

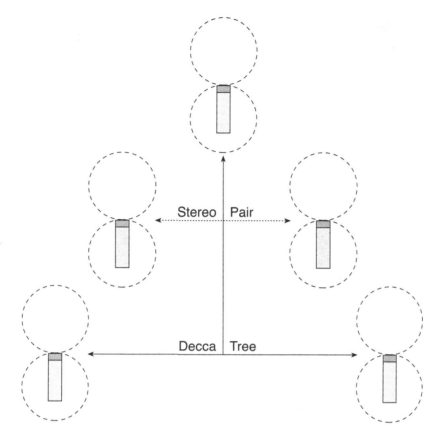

Figure 14.2 Decca tree and A-B stereo pair for solo cello.

DOUBLE BASS AND PIANO

Jeff Wolpert (Desert Fish Studios) recorded Joel Quarrington's and David Jalpert's album *An die Musik* on 5 and 6 December 2016 at the Isabel Bader Centre for the Performing Arts in the acoustically isolated 566-seat performance hall at the heart of the building (Queen's University, Canada).[3] The architects designed the stage, located at one end of the rectangular space, as an area open from above but circumscribed by back and side walls (see Figure 14.3).

To capture the room's reverberation in a natural way that suits double bass and piano, Wolpert angled the piano towards the rear of the stage, with the back of the open lid facing the audience. Quarrington stood to Jalpert's left, beside the piano, as this allowed the double bass to point more-or-less directly into the hall. A baffle between the instruments helped reduce spill, and an ORTF array (Neumann KM 84 cardioid mics) centered on the curve of the piano, just outside the case, picked up direct sound from the instrument, as well as the first reflections from the nearby walls. A Faulkner phased array in front of the double bass, with the nulls of the two ribbon mics (Beyerdynamic M 130) directed at

Figure 14.3 Performance Hall, Isabel Bader Centre for the Performing Arts, Queen's University (Kingston).

Source: Photograph: Suzy Lamont Photography. Used with the permission of Suzy Lamont Photography and Queen's University (Kingston).

the piano, provided the main soloist capture, which was supplemented by three spot microphones (AKG C480s with CK62 omni capsules), one at the back and one on each side of the double bass. This combination of mics let Wolpert balance the amount and quality of the bass presence with that of the piano, and the resulting recording sounds spacious, yet intimate. In fact, the ratio of direct to ambient sound achieved through Wolpert's approach allows the double bass and piano to interact in a transparent manner.

NOTES

1. See the postings from 2 and 3 December 2008 at www.gearslutz.com/board/remote-possibilities-in-acoustic-music-and-location-recording/333838-whats-up-classical-piano-recordings-these-days-3.html. I am grateful to Simon Eadon for confirming the details posted on the site. The recordings appear on Hyperion, CDA67569, 2007.
2. The information in this section has been taken from an educational video on the website of Royer Labs (http://royerlabs.com/library/winona-zelenka/). The recordings are available from Marquis Classics, MAR 509. I am grateful to Ron Searles and Winona Zelenka for clarification of the details in this section.
3. The recording is available on iTunes and at https://joelquarrington.com/store/schubert-an-die-musik-physical-cd. I am grateful to Jeff Wolpert for his willingness to share his recording strategies with me.

REFERENCES

Hamelin, Marc-André. 2007. *Troisième recueil de chants*. London: Hyperion Records.

Keller, Nicolas. 2013. "Behind the Sessions, Recording Classical Piano with the Focusrite ISA MKII at the IOSR." 10 January. Available at: www.youtube.com/watch?v=e2t91591ISk.

Quarrington, Joel and David Jalbert. 2017. *An die Musik*. iTunes Store. Available at: https://joelquarrington.com/store/schubert-an-die-musik-physical-cd.

Zelenka, Winona. n.d. *J.S. Bach: Six Suites for Solo Cello*. Toronto, Canada: Marquis Classics.

CHAPTER 15

Studio Techniques

Re-Creating the Aural Sense of Historic Spaces

Musicians who wish to take a historically informed approach to performing music composed between the sixteenth and nineteenth centuries often find themselves without access to a suitable room for tracking, and in the absence of acoustically appropriate historic spaces, large churches have become favored venues for recording early music, even though they are far too reverberant for much of the repertoire, particularly solo songs accompanied by quiet instruments, such as the lute, guitar, or harpsichord.[1] Modern studio technology, however, can be employed to simulate the aural sense of the modest chambers in which the music probably would have been performed originally,[2] especially convolution reverbs based on impulse responses taken from smaller rooms in seventeenth- or eighteenth-century buildings and artificial reverberation of the reflection-simulation type designed to mimic the characteristics of generic spaces.

This chapter focuses on the approach a group of recordists took not only to achieve a period-style interpretation of an early eighteenth-century cantata but also to locate that performance in an acoustic designed to give listeners the impression that they are sitting in the same small room as the performers. Specifically, I discuss a track from Studio Rhetorica's album *Secret Fires of Love* (2017) that features tenor Daniel Thomson and harpsichordist Thomas Leininger under my musical direction. The recording of Tomaso Albinoni's "Amor, sorte, destino" (*12 Cantate da camera a voce sola*, op. 4, Venice, 1702) was produced by me and recorded and mixed by Robert Nation (Kyle Ashbourne, assistant engineer) at EMAC Recording Studios in London, Canada (the track may be downloaded at CD Baby or streamed at Apple Music and Spotify, among others). I am most grateful to Robert for discussing his philosophies/procedures of recording and mixing with me, as the production and post-production sections below blend explanations of his practices with my contributions as music director and producer.

The first part of the chapter focuses on the pre-production stage of the project, during which Daniel, Thomas, and I finalized the interpretive strategies that would be employed, and the following sections concentrate on the ways studio production practices at EMAC aided the transference of a historically informed conception of "Amor, sorte, destino" to disk.

PRE-PRODUCTION

Older principles of interpretation differ considerably from those currently used by classical musicians, and in order for people interested in historical performance to recover the old methods, they must reconstruct the practices from surviving sources of information. Fortunately, a great deal of material comes down to us, and this allowed us not only to root the interpretive strategies employed in *Secret Fires of Love* in period documents but also to take a fresh approach to Baroque vocal works. Recent research has shown that from the sixteenth to the early nineteenth centuries, singers modeled their art directly on oration and treated the texts before them freely to transform inexpressive notation into passionate musical declamation (see, in particular, Toft 2013, 2014).

Daniel adopts the persona of a storyteller, and like singers of the past, he uses techniques of rhetorical delivery to re-create the natural style of performance listeners from the era probably would have heard. This requires him to alter the written scores substantially, and his dramatic singing combines rhetoric and music in ways that sympathetically resonate with performance traditions from the Baroque era. In "Amor, sorte, destino," Daniel treats the two recitatives differently from the two arias; sings prosodically, emphasizing important words and giving the appropriate weight to accented and unaccented syllables; employs a highly articulated manner of phrasing; alters tempo frequently through rhythmic *rubato* and the quickening and slowing of the overall time; restores *messa di voce*, the swelling and diminishing of individual notes, as well as phrases, to its rightful place as the "soul of music" (Corri 1810: I, 14); contrasts the tonal qualities of chest and head voice as part of his expression; and applies *portamento*.

Among these principles, highly articulated phrasing, alterations of tempo, and variations in the tonal quality of the voice represent the most noticeable departures from modern practice. Singers of the past inserted grammatical and rhetorical pauses to compartmentalize thoughts and emotions into easily discernible units (that is, stops at punctuation marks and in places where the sense of the sentence called for them), and this frequent pausing gave listeners time to reflect on what they had just heard so they could readily grasp the changing sentence structure. As early as 1587, Francis Clement had explained the rationale behind the addition of unnotated pauses: "the breath is relieved, the meaning conceived, . . . the eare delited, and all the senses satisfied" (Clement 1587: 24–5; for further information on pausing, see Toft 2013: 20–45; Toft 2014: 84–98). Moreover, writers from Nicola Vicentino (1555: 94v) to Giambattista Mancini (1774: 150) observe that singers best convey the true sense and meaning of words in a natural way if they derive the pacing of their delivery from the emotions in each text segment. Or to use Vicentino's words (1555: 94v), tempo fluidity has "a great effect on the soul/effetto assai nell'animo" (tempo modification is discussed further below).

Similarly, the use of appropriate vocal timbres to carry the text's emotions to the ears of listeners requires singers not only to differentiate their registers (so that the lowest and highest parts of the range contrast with the middle portion) but also to link timbre and emotion (smooth and sweet, thin and choked, harsh and rough)—"the greater the passion is, the less musical will be the voice that expresses it" (Anfossi c.1840: 69). In earlier eras, a versatile tonal palette prevented the monotony of what David Ffrangcon-Davies dismissed in 1905 as the "school of sensuously pretty voice-production." Indeed, as Ffrangcon-Davies suggests, the then new monochromatic approach to timbre meant that if audiences had heard a singer in one role, they had heard that singer in every role (1905: 14–16).

Armed with a collection of historic principles, our first task in re-creating period style was to study the lyrics of "Amor, sorte, destino" to find all the places a singer might wish to insert grammatical or rhetorical pauses. This exercise involved following principles described in treatises on rhetoric and oration to add pauses of varying lengths at points of punctuation (grammatical pauses) and to separate subjects from verbs and verbs from objects, as well as conjunctions, relative pronouns, participles, and prepositional phrases from the text that precedes them (rhetorical pauses). The compartmentalization of ideas and emotions organizes and paces the content of the poem so that listeners can easily grasp the story, and since some ideas require a slower or quicker delivery than others, compartmentalization also provides appropriate places for singers to change the speed of delivery to match the emotional character of the phrases.

But apart from organizing and pacing ideas and emotions, singers must decide which word or words within a phrase should be emphasized, and treatises usually combine the discussion of emphasis with that of accent. Accent denotes the stress placed on a single syllable to distinguish it from the others in a word (this is known as speaking or singing prosodically), whereas emphasis refers to the force of voice laid on an entire word or group of words to bring the associated ideas to the attention of listeners. Proper accentuation, then, adhered to the normal pronunciation of words in ordinary speech, while emphatic delivery varied according to the meaning performers wished to convey (for more information on accent and emphasis, see Toft 2013: 73–9; Toft 2014: 98–108).

Emphatic words, then, receive the greatest force within a sentence, and these important words are situated in an overall hierarchy of emphasis in which speakers reserve the strongest sound of voice for the most significant word or idea, firmly and distinctly pronouncing substantives (nouns), adjectives, and verbs, while relegating unimportant words (the, a, to, on, in, of, by, from, for, and, but, etc.) to relative obscurity (Walker 1781: II, 15, 25). This hierarchy allows performers not only to arrange words into their proper class of importance but also to achieve a distribution of emphases that would prevent sentences from being delivered monotonously with uniform energy (Herries 1773: 218). Thus, the application of accent and emphasis creates light and shade and helps speakers (and singers) clearly project the meaning of long and complex ideas. Inappropriate shading would force listeners to decipher a sentence's meaning from an ambiguous or confusing delivery (Murray 1795: 153).

After we completed our analysis of the song's text using the principles just discussed, Daniel proceeded to create a dramatic spoken reading of the poem.

Initially, this meant deciding which pauses would be employed and what ideas would be exhibited prominently (that is, emphasized), as well as what variations in the speed of delivery would suit the changing emotions of the text. As part of this process, Daniel also took note of where *messa di voce* and *portamento* occurred as he spoke, for in these places both the swelling and diminishing of the voice and the sliding between pitches would sound the most natural in singing (*messa di voce* and *portamento* are discussed more fully in Toft 2013: 45–69). Once we were satisfied with the spoken narrative, Daniel transferred his interpretation to the song, altering Albinoni's melodic lines to accommodate the dramatic reading.

By rooting our performance in historical documents, we were able to model our understanding of the relationship between performer and score directly on principles from the past. Indeed, singers in earlier times viewed scores quite differently from their modern counterparts. They realized that because composers wrote out their songs skeletally, performers could not read the notation literally, and to transform inexpressively written compositions into passionate declamation, vocalists treated texts freely and personalized songs through both minor and major modifications. In other words, singers saw their role more as one of re-creation than of simple interpretation, and since the final shaping of the music was their responsibility, the arias and recitatives listeners heard often differed substantially from what appeared in print (Toft [2013: 4–6] discusses the relationship between notation and performance).

Composers of the past did not notate subtleties of rhythm, phrasing, dynamics, pauses, accents, emphases, tempo changes, or ornamentation. Clearly, they had no desire (or need) to capture on paper the elements of performance that moved listeners in the ways writers from the time described. In the middle of the sixteenth century, Nicola Vicentino (1555: 94v) commented that "sometimes [singers] use a certain method of proceeding in compositions that cannot be written down"/"*qualche volta si usa un certo ordine di procedere, nelle compositioni, che non si può scrivere,*" and along these lines, Andreas Ornithoparchus, writing in 1517 (p. 89 in John Dowland's 1609 translation), praised singers in the Church of Prague for making "the Notes sometimes longer, sometime[s] shorter, then they should." Around 1781, Domenico Corri (i, 2) characterized the relationship between performance and notation candidly: "either an air, or recitative, sung exactly as it is commonly noted, would be a very inexpressive, nay, a very uncouth performance." Charles Avison had already made this notion explicit in 1753 (p. 124): "the Composer will always be subject to a Necessity of leaving great Latitude to the Performer; who, nevertheless, may be greatly assisted therein, by his Perception of the Powers of Expression," and a hundred years later voice teachers like Manuel García (1857: 56) continued to suggest the same thing—performers should alter pieces to enhance their effect or make them suitable to the power and character of an individual singer's vocal capability.

As far back as 1555, Nicola Vicentino (fol. 94v) suggested why performers valued flexibility of tempo:

The experience of the orator teaches this [the value of changing tempo (*mutar misura*) within a song], for one sees how he proceeds in an oration—for now he speaks loudly and now softly, and more slowly and more quickly, and with this greatly moves his auditors; and this way of changing the tempo has a great effect on the soul. [*La esperienza, dell'Oratore l'insegna, che si vede il modo che tiene nell'Oratione, che hora dice forte, & hora piano, & più tardo, & più presto, e con questo muove assai gl'oditori, & questo modo di muovere la misura, fà effetto assai nell'animo.*]

Hence, vocalists sang *piano e forte* and *presto e tardo* not only to conform to the ideas of the composer but also to impress on listeners the emotions of the words and harmony. In the early nineteenth century, Domenico Corri (1810: I, 6) suggested a similar approach under the heading "Quickening or retarding of time:"

> Another improvement, by deviation from strict time, is to be made by the singer delivering some phrases or passages in quicker or slower time than he began with, in order to give emphasis, energy, or pathos, to particular words.

Primarily, the period-specific alterations we made to Albinoni's text and melodic lines involved adding pauses and adjusting the rhythmic values of the notes so that the delivery of the syllables and words came as close to speaking as possible. But Daniel also varied tempo along the lines Vicentino and Corri had suggested and employed light and shade (accent and emphasis) in a historic way, for if he were to sing the melodies exactly as Albinoni had notated them, he would, in Corri's view, be guilty of an "inexpressive" and "uncouth" performance. By placing his persona as a storyteller in this older guise, Daniel has provided what one writer, John Addison (c.1850: 29), called the "finish" to the song in a way that approximates early eighteenth-century style.

PRODUCTION

The main goal in producing "Amor, sorte, destino," as well as the other tracks on *Secret Fires of Love*, was to enhance period interpretation through modern studio practices, especially isolated sound sources recorded by closely placed microphones. We chose to blend the worlds of recording and "live" performance so that we could capture a dramatic reading of the song, while achieving sonic clarity. In other words, we did not consider the two activities to be mutually exclusive, and because we felt that making records and archiving a "live" event differed fundamentally, we decided to use punch-ins to perfect excellent takes rather than completely re-record those sections that contained minor imperfections.

From these perspectives, the project benefited from having one person assume the roles of music director and producer, for a single conception of the cantata could then emerge from the various sonic possibilities available to the artists and engineers. Indeed, decisions made throughout the process, from those that led to a historically relevant interpretation of the printed score to those that guided the design of the soundscape in which the performance was presented, came from imagining how one world might inform the other. Knowledge of both historical performance and recording practices focused the energies of everyone involved in the project on an idealized conception, and the various elements of production, when combined with sympathetic strategies for editing and mixing, helped shape the recording along historical lines.

Robert Nation tracked in Pro Tools HD at 24 bits/96 kHz to provide an excellent signal-to-noise ratio, as well as increased dynamic range, and microphone selection and placement figured prominently at the beginning of the sessions. Since harpsichords can be somewhat bright in tonal quality, the question arose of how we might best capture the sound of the instrument. The mics chosen would need to produce as "natural" a stereo sound as possible when positioned closely, so microphones with too much boost on the top end would not be suitable, as they would exaggerate the brightness of the instrument. In fact, a stereo pair of omnidirectional microphones with a linear frequency response would probably be ideal for this application, as omnis would allow the characteristic resonance of the instrument to be portrayed realistically, without proximity effect. For the voice, a microphone that could provide a consistent frequency response across Daniel's range, while keeping sibilance to a minimum, would be preferable, and both ribbon and omnidirectional microphones were obvious possibilities.

A "shoot-out" using mics from EMAC's and my collections resulted in the following choices:

Voice—Royer R-122 active ribbon mic (a Schoeps small-diaphragm condenser with an MK 2 capsule, omnidirectional with a flat frequency response, was also quite attractive)

Harpsichord—for close mics, a stereo pair of DPA's omnidirectional 4006A, and for room mics, stereo pairs of Milab's multi-pattern DC-196 and AKG's omnidirectional 480B.

Daniel sang in an isolation booth with his microphone, shielded by a pop screen, placed approximately 30 centimeters (12 inches) in front of his mouth. Although ribbon mics exhibit a fairly strong proximity effect, the Royer R-122 not only provided the consistency of frequency response we desired but also eliminated most of the problems associated with sibilance. Thomas performed in the main tracking room, and because the dynamic range of harpsichords is not large, the first pair of mics for the instrument (DPA 4006A) were placed under the lid in an A-B arrangement spread bass to treble to provide a natural tonal balance. A second pair of stereo mics, Milab DC-196s in ORTF configuration, were positioned a short distance from the harpsichord along its center line, and a third pair of mics, AKG 480Bs mounted on either side of a Jecklin disk, contributed additional room perspective from slightly farther away.

Once appropriate levels had been set, the tracking procedure consisted of several initial takes of the whole cantata. The entire team then listened to these recordings to determine if one of them could become a master take that would be refined through punch-ins or if we should record the cantata movement by movement. The group decided to work on the two recitatives separately from the two arias, and after recording several takes of each section, the team chose which recordings would comprise the master track. Both Daniel and Thomas then punched in minor corrections, before the assembled material was deemed ready for editing and mixing.

POST-PRODUCTION

The editing process not only helped us achieve our ideal historically informed conception of "Amor, sorte, destino" but also allowed us to reduce any noise in the recording that might distract listeners. Because audio recordings lack the visual connection of "live" performance, extraneous sounds that mar the sensory surface of a disk can be much more disruptive, especially jack noise from harpsichords and vocal plosives. Kyle Ashbourne, the assistant engineer, and I carefully listened to the master track and used the spectral repair feature of iZotope's *Rx Advanced* to decrease jack noise to an amount a listener would hear when seated 3–4 meters (10–12 feet) away from the performers. In other words, we wished to present the harpsichord from the perspective of the listener rather than the player.

Similarly, plosives that resulted from the singer's closely placed ribbon microphone were reduced using the high-pass filter in Universal Audio's *Massenburg DesignWorks MDWEQ5*. Moreover, Pro Tools' clip gain helped us achieve the prosodic manner of vocal delivery prized in the past, for we lowered the level of those syllables that close miking had heightened. In addition, whenever the Royer R-122 exaggerated a *messa di voce* with too much energy at its peak, automation of the dynamics in Pro Tools helped bring those phrases into line with the way they would be heard a short distance from a singer (here we were thinking of how the Inverse Square Law affects the propagation of the soundwaves to soften the swells naturally).

The mixing sessions focused on three main elements—reverb, compression, and EQ. As we began to consider possible models for creating a suitable ambience, we decided to listen to a number of harpsichord performances that had been recorded in large churches, the typical locations for such recordings. These rooms were, of course, far too reverberant for our purposes, and we realized that without appropriate models to emulate we needed to imagine a performance space that did not yet exist on a recording. Robert then set about designing an artificial ambience that would approximate the small rooms in which the music was often performed originally.

He chose the same two algorithmic reverbs for the voice and the harpsichord, so that Daniel and Thomas would sound like they were performing together in a room. The first was the Large Hall B in Universal Audio's *Lexicon 224 Digital Reverb*, with a decay time of 2.0 sec in the vocal and 1.7 sec in the harpsichord, and the second was the Small Plate 2 in Eventide Audio's *UltraReverb*, set to a short decay time of 837 ms to "tighten" the ambience surrounding both performers. Robert used the two reverbs quite subtly, mixing them in at a low level, and because Daniel has a large dynamic range, the input to the *Lexicon 224* was compressed to bring the level of the loudest passages down by 2.0–3.0 dB, so that Daniel's voice would not over-trigger the reverb. Compression, then, helped make the vocal ambience less obvious. Harpsichords, on the other hand, have quite a small dynamic range, and since they could never over-trigger the *Lexicon 224*, Thomas' input did not need to be compressed.

Beyond the reflection simulation Robert crafted to place the performers in an appropriately sized room, he included two parallel effects busses on the vocal track and adjusted the frequency balance of Daniel's voice. A compression back bus, combining Universal Audio's *EL7 FATSO Sr* with Sonnox' *Oxford Inflator*, was mixed in at a very low level (–24.0 dB) to increase the intimacy of the quietest passages, and on another bus, Nugen Audio's *Stereoizer* created a bit of extra space in the mix. Specifically, delays of 12 ms on one side and 19 ms on the other added the sense of early reflections which, together with the reverbs, compensated for recording into a baffle.

To lessen the proximity effect inherent in the Royer R-122, the *Massenburg DesignWorks MDWEQ5* parametrically removed some of the energy around 310 Hz, while the high-pass filter of the same plugin disposed of any rumble below 40 Hz. Close miking of a singer can also produce some mild harshness at the loudest moments, and a cut of 1.4 dB in Brainworx' dynamic equalizer *bx_dynEQ V2* was used to alleviate this tension around 2,357 Hz.

Robert also adjusted the frequency balance in the harpsichord tracks, and he employed the high-pass filter in McDSP's *FilterBank F202* to eliminate any rumble below 91.42 Hz in the microphone channels, while placing an extra high-pass filter (Nugen Audio's *SEQ-ST*) in the channel strip for the close mics to remove some noticeable "thumping" the DPAs had picked up below 55 Hz. In addition to these adjustments, he enhanced the overall character of the harpsichord with the "Dark Essence" setting in Crane Song's tape-emulation plugin *Phoenix II*, which gave the instrument a fuller sound.

On the master bus, because downstream codecs can increase the peak level of the signal somewhat, a true-peak limiter (Nugen Audio's *ISL 2*) was set at –0.7 to leave room for file conversion, and the general loudness characteristics of the track were analyzed through Nugen Audio's *MasterCheck*.

* * * * *

In our recording of "Amor, sorte, destino," as well as in the other tracks on *Secret Fires of Love*, we clearly embraced the notion that "the most important reverberation exists within the recording, not the playback space" (Case 2007: 263), and since we did not want listeners to experience the music from a distance, as if they were in a large church or concert hall, we decided to create an artificial ambience that would situate them about 3–4 meters (10–12 feet) from the artists. Hence, a blend of close miking and the digital processes described above, all in the service of a historically informed performance, allowed everyone involved in the project (artists, producer, and engineers) to realize on disk what we imagined someone in the eighteenth century might have heard in the small rooms in which the music was frequently performed.

NOTES

1. Recent research has shown that musicians in the sixteenth to eighteenth centuries regularly performed in private chambers or small music rooms (Howard & Moretti 2012: 106–7, 111–14, 185–9, 200–3, 248–9, 320), and because many of the *camere per musica* measured no more than 7 by 11 meters (23 by 36 feet), with a ceiling height of 5 to 6 meters (16 to 20 feet), their volumes (approximately 385 cubic meters or 13,250 cubic feet) produced reverberation times of less than a second, which means that these spaces tended more to clarity and intimacy than reverberance (Orlowski 2012: 157–8). Moreover, since listeners would have been seated close to the performers in these rooms, direct sound would predominate, and early reflections from the walls and floor would further contribute to the sonic impression of clarity and intimacy (Orlowski 2012: 158).
2. For drawings of the modest rooms in which Giulio Caccini most likely sang his own compositions in the late sixteenth and early seventeenth centuries, see Markham 2012: 200–3.

REFERENCES

Addison, John. c.1850. *Singing, Practically Treated in a Series of Instructions*. London: D'Almaine.

Anfossi, Maria. c.1840. *Trattato teorico-pratico sull'arte del canto . . . A Theoretical and Practical Treatise on the Art of Singing*. London: By the Author.

Avison, Charles. 1753. *An Essay on Musical Expression*. London: C. Davis. Reprint, New York, NY: Broude, 1967.

Case, Alexander U. 2007. *Sound FX: Unlocking the Creative Potential of Recording Studio Effects*. Boston, MA: Focal Press.

Clement, Francis. 1587. *The Petie Schole*. London: Thomas Vautrollier. Reprint, Leeds, UK: Scolar, 1967.

Corri, Domenico. c.1781. *A Select Collection of the Most Admired Songs, Duetts, &c.* 3 vols. Edinburgh, Scotland: John Corri. Reprint, Richard Maunder. *Domenico Corri's Treatises on Singing*. Vol. 1. New York, NY: Garland, 1993.

Corri, Domenico. 1810. *The Singer's Preceptor*. London: Longman, Hurst, Ress, and Orme. Reprint, Richard Maunder. *Domenico Corri's Treatises on Singing*. Vol. 3. New York, NY: Garland, 1995.

Ffrangcon-Davies, David. 1905. *The Singing of the Future*. London: John Lane.

García (the Younger), Manuel. 1857. *New Treatise on the Art of Singing*. London: Cramer, Beale, and Chappell.

Herries, John. 1773. *The Elements of Speech*. London: E. and C. Dilly. Reprint, Menston, UK: Scolar, 1968.

Howard, Deborah and Laura Moretti, eds. 2012. *The Music Room in Early Modern France and Italy: Sound, Space, and Object*. Oxford, UK: Oxford University Press for The British Academy.

Mancini, Giambattista. 1774. *Pensieri, e riflessioni pratiche sopra il canto figurato*. Vienna, Austria: Stamparia di Ghelen.

Markham, Michael. 2012. "Caccini's Stages: Identity and Performance Space in the Late Cinquecento Court." In Howard and Moretti 2012: 195–210.

Murray, Lindley. 1795. *English Grammar*. York: Wilson, Spence, and Mawman. Reprint, Menston, UK: Scolar, 1968.

Orlowski, Raf. 2012. "Assessing the Acoustic Performance of Small Music Rooms: A Short Introduction." In Howard and Moretti 2012: 157–9.

Ornithoparchus, Andreas. 1517. *Musice active micrologus*. Leipzig, Germany: Valentin Schumann. Translated John Dowland. *Andreas Ornithoparcus His Micrologus, or Introduction*. London: Thomas Adams, 1609. Reprint of both editions in one volume, New York, NY: Dover, 1973.

Studio Rhetorica. 2017. *Secret Fires of Love*. London: Talbot Records.

Toft, Robert. 2013. *Bel Canto: A Performer's Guide*. New York, NY: Oxford University Press.

Toft, Robert. 2014. *With Passionate Voice: Re-Creative Singing in Sixteenth-Century England and Italy*. New York, NY: Oxford University Press.

Vicentino, Nicola. 1555. *L'antica musica ridotta alla moderna prattica*. Rome, Italy: Antonio Barre. Reprint, Kassel, Germany: Bärenreiter, 1959.

Walker, John. 1781. *Elements of Elocution*. London: T. Cadell.

Glossary

"Active" device—one that requires external power to function.

Aliasing—sampling rates that are too low to map a waveform accurately prohibit the faithful restoration of signals. A fault known as aliasing occurs when too few samples cause a device to interpret the voltage data as a waveform different from the one originally sampled.

Amplitude—a measure of the change of air pressure in a soundwave above normal (compression) and below normal (rarefaction). In other words, it is a measure of the strength of a sound without reference to its frequency. We perceive amplitude as loudness and express it in decibels (dB) of sound pressure level (SPL).

Analog audio—the representation of a signal by continuously variable and measurable physical quantities, such as pressure or voltage.

Analog-to-digital converter (ADC)—converts analog signals to digital code using pulse code modulation (PCM).

Audio volume—in relation to the measurement of loudness, "audio volume" refers to a subjective combination of level, frequency, content, and duration.

Bit—an abbreviation of the expression "binary digit." Binary means something based on or made up of two things, and in digital audio systems, these two things are the numbers 0 and 1.

Bit depth—stipulates how many numbers (0 or 1) are used to represent each sample of a waveform.

Bit rate—indicates how many bits are transmitted per unit of time in digital audio.

Byte—a group of eight digits, each digit being either 0 or 1.

Codec—an abbreviation of coder/decoder, a codec is a software application using algorithms to encode a digital signal into another format, often to reduce the size of the file (hence the term compression). Once encoded, the file must be decoded to re-create the original audio. If no information is lost in the process, the codec is called lossless, but if information has been removed, the codec is a lossy one.

Coloration—an audible change in the quality (timbre) of a sound.

Comb filtering—the short delays between two or more microphones used to capture a complex waveform can cause comb filtering, a set of mathematically related (and regularly recurring) cancellations and reinforcements in which the summed wave that results from the inadvertent cutting and boosting of frequencies resembles the teeth of a comb.

Complex soundwave—a waveform comprised of a collection of sine waves, integer multiples of the fundamental frequency, that is, a complex set of frequencies arranged in a harmonic or overtone series above the lowest frequency of the spectrum.

Condenser microphone—a mic that operates electrostatically. Its capsule consists of a movable diaphragm and a fixed backplate, which form the two electrodes of a capacitor (previously called a condenser; hence, the name) that has been given a constant charge of DC voltage by an external power source. As soundwaves strike the diaphragm, the distance between the two surfaces changes, and this movement causes the charge-carrying ability (capacitance) of the structure to fluctuate around its fixed value. The resulting variation in voltage creates an electrical current that corresponds to the acoustic soundwave.

Convolution—the blending or convolving of one signal with another. Convolution is the method used to create reverb plugins based on impulse responses.

Critical distance (reverberation radius)—the distance from a sound source to that point in an enclosed space where the direct and reverberant fields are equal in level; that is, the total energy of one equals the other. Physicists have determined that the level of direct sound drops by 6.0

dB for every doubling of distance in a truly free field (that is, outdoors; the drop is somewhat smaller in an enclosed space) and that the level of reverberated sound remains more or less constant everywhere in a room. The ratio of direct to reverberated sound is 1:1 at the critical distance (see Figure Glossary.1).

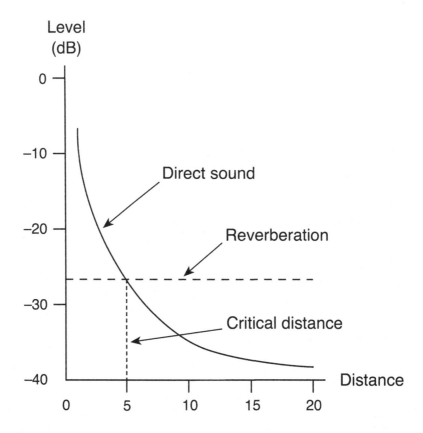

Figure Glossary.1 Critical distance.

The critical distance may be found in any room by at least two methods: (1) place a microphone relatively far from a sound source and then move a second mic increasingly closer to the source until the difference in level between the two microphones is less than 3.0 dB; (2) either during a rehearsal or by situating a boom box at the performers' location (set between stations to produce white noise), recordists use an SPL meter to measure the SPL close to the source (approximately 30 centimeters or a foot away) and then double the distance and measure again, a point at which, according to the Inverse Square Law, the level will have decreased between 4.0 and 6.0 dB. After making note of the new level, they double the distance and take another measurement, repeating this procedure until the SPL stops dropping. By moving back to the area where the level began to remain constant, they find the critical distance.

At distances less than a third of the reverberation radius, the direct sound will be at least 10.0 dB stronger than the reverberated sound; hence, reverberation does not play a prominent role in the sound captured by a microphone. Conversely, at distances three times that of the critical distance, the direct sound is at least 10.0 dB weaker than the reverberated sound, and a microphone will primarily capture reverberation.

Damping—a method of controlling the way frequencies die away or roll off in a reverb tail created during digital reflection simulation.

dBFS (dB Full Scale)—audio level in decibels referenced to digital full scale, that is, referenced to the clipping point ("full scale") in a digital audio system. 0.0 dB represents the maximum level a signal may attain before it incurs clipping.

dBTP (dB True Peak)—maximum inter-sample peak level of an audio signal in decibels referenced to digital full scale, that is, referenced to the clipping point ("full scale") in a digital audio system. 0.0 dB represents the maximum level a signal may attain before it incurs clipping.

Decibel (dB)—one tenth of a bel. Named after Alexander Graham Bell, a bel expresses the logarithmic relationship between any two powers. In acoustics, large changes in measurable physical parameters (pressure, power, voltage) correspond to relatively small changes in perceived loudness. Thus, linear scales, because of the huge numbers involved, do not correspond very well to the perceived sound, so a logarithmic scale is used to bring the numerical representation of perceived loudness and the numerical representation of the actual physical change into line with each other. Logarithms are a simple way of expressing parameters that vary by enormous amounts with smaller numbers (in other words, a large measurement range is scaled down to a much smaller and more easily usable range).

Because the human ear accommodates a large range of loudness, it is convenient to express loudness logarithmically in factors of ten. The entire range of loudness can be expressed on a scale of about 120.0 dB (0.0 dB is defined as the threshold of hearing), and within this logarithmic scale, increasing the intensity of sound by a factor of 10 raises its level by 10.0 dB, increasing it by a factor of 100 raises the level by 20.0 dB, and increasing it by a factor of 1,000 raises the level by 30.0 dB, and so on. Hence, the term decibel does not represent a physical value. It is a relative measurement based on the internationally accepted standard that 20 micropascals of air pressure equals 0.0 dB (20 micropascals of air pressure at 1,000 Hz is the threshold of hearing for most people).

Since the term decibel expresses a ratio and not a physical value, it can be applied to things other than loudness. In amplifiers, for example, a 200 watt amp is 1 bel or 10 decibels more powerful than a 20 watt amp, but it is important to understand that even though a 200 watt amp puts out ten times more electrical power than a 20 watt amp, it does not generate ten times more loudness, for a ten-fold change of electrical power is only perceived by the human ear as a 10.0 dB change of loudness. In other words, the underlying scales are different, and one scale should not be equated directly with the other. This can be shown in a graph, where the vertical axis represents dB and the horizontal axis represents electrical power (the curved line in Figure Glossary.2 is the logarithmic contour).

Diaphragm—the thin membrane in a microphone capsule that moves in reaction to soundwaves. In the early days of capacitor microphones, diaphragms were made from PVC (polyvinyl chloride, such as in the M7 capsule designed by Georg Neumann in 1952), but now they are usually made from PE (polyethylene), which is lighter, thus providing a more responsive capsule with greater sensitivity and articulation. Manufacturers fashion these materials into thin sheets coated with a gold surface so that the diaphragm may be charged to create a capacitive effect (hence, the term "metal film"). In the most expensive capsules, the gold is evaporated onto the membrane in a vacuum chamber to ensure uniform coverage. The more economical process of sputtering or spraying gold onto the membrane, the process used on less expensive microphones, can result in an uneven coat that causes membrane imbalance and inconsistencies in the capsule's response. In ribbon mics, the diaphragm consists of a thin strip of corrugated aluminum.

Diffuse or reverberant sound field—the area in a room in which reflections from the walls, ceiling, floor, etc. predominate (that is, the ensemble of reflections in an enclosed space). In other words, the sounds arrive at the listening position/microphone randomly from all directions and the direct sound no longer dominates. These reflections have their high-frequency content

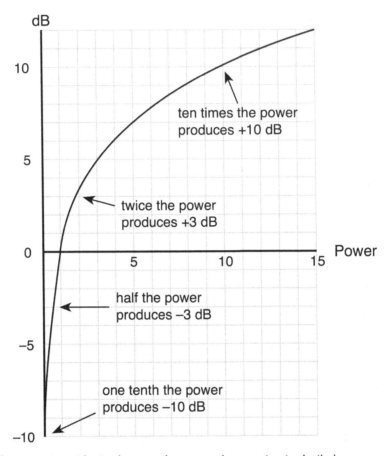

Figure Glossary.2 Logarithmic changes of power and perception in decibels.

attenuated by surface absorption, as well as by the air, and reach the listener/microphone at oblique angles of incidence, which causes further high-frequency loss. This is also the field in which direct sound travels as a plane wave (in plane waves, intensity decreases in a linear relationship to distance traveled, that is, the Inverse Square Law no longer applies).

Digital audio—the use of a series of discrete binary numbers (0 or 1) to represent the changing voltage in an analog signal.

Digital-to-analog converter (DAC)—converts digital code to an analog signal (voltage), so that non-digital systems can use the information.

Direct or free sound field—sound arriving perpendicularly to the listening position or the diaphragm of a microphone without reflections (a purely direct field can exist only where sound propagation is undisturbed, such as in an open space free from all reflections). The direct path of sound is the shortest route from the sound source to the listening position. It is also the area in which soundwaves propagate spherically and the Inverse Square Law applies. This field ends where the sound pressure level ceases to fall by 6.0 dB for every doubling of the distance.

Distance factor—an indication of how far recordists can locate a directional microphone from a sound source and have it exhibit the same ratio of direct-to-reverberant sound pickup as an omnidirectional microphone.

Dither—the technique of adding specially constructed noise to a signal before its bit depth is reduced to a lower level. It alleviates the negative effects of quantization by replacing nonrandom distortion with a far more pleasing random noise spectrum. One of the commonly used types of dither is TPDF (triangular probability density function, which uses white noise with a flat frequency spectrum), but devices can also add noise containing a greater amount of high-frequency content (called blue noise). The process involving blue noise is known as colored/shaped noise dithering or noise shaping, and it concentrates the noise in less audible frequencies (generally those above 15–16 kHz), while reducing the level of the noise in the frequency range humans hear best (between 2 and 5 kHz and around 12 kHz).

Ducking—the technique of dropping one signal below another. It is frequently used in voiceovers to place the main signal in the background while the announcer speaks.

Dynamic microphone—these microphones operate on the principle of electromagnetic induction. A light diaphragm connected to a finely wrapped coil of wire suspended in a magnetic field moves within that magnetic field to induce an electrical current proportional to the displacement velocity of the diaphragm. Dynamic microphones are also called velocity or moving-coil microphones.

Dynamic range—the difference between the softest and loudest sound a system can produce.

Early reflections—the first reflections to arrive at a listening position within 80 ms of the direct sound.

Equal loudness curves—these curves or contours, originally established by the researchers Harvey Fletcher and Wilden A. Munson in the 1930s (and refined by later researchers), show how loudness affects the way humans hear various frequencies. Figure Glossary.3 demonstrates that people exhibit the greatest sensitivity to frequencies around 4 kHz and the least sensitivity at either end of the spectrum, particularly in the lower part of the hearing range. In other words, for listeners to perceive a 50 Hz sound in the same way they perceive a 1 kHz sound at 40.0 dB, the level of the signal has to be increased to 70.0 dB (in the chart, follow the 40 phon contour from 1 kHz up to the 70.0 dB level and the frequency below this point is roughly 50 Hz).

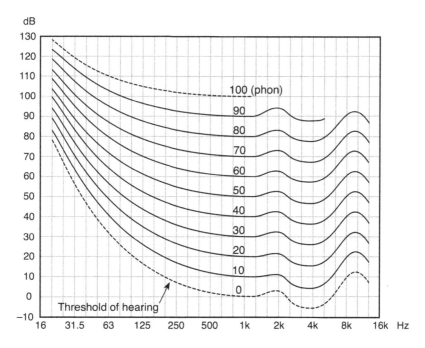

Figure Glossary.3 Equal loudness curves (contours).

Fast Fourier transform (FFT)—a type of mathematical analysis, first developed by Jean Baptiste Joseph Fourier in the early nineteenth century, that allows data from one domain to be transformed into another domain. Computers perform the Fourier transform (that is, the mathematical calculations for it) at a very high speed, and modern spectrograms rely on what is known as the "fast Fourier transform" to plot frequency against amplitude in real time so that the visual representation of a signal changes as rapidly as the signal itself.

Filter—any device that alters the frequency spectrum of a signal by allowing some frequencies to pass, while attenuating others. Filters change the balance between the various sine waves that constitute a complex waveform.

Free sound field—*see* Direct or free sound field.

Frequency—a measure of how often ("frequently") an event repeats itself. A sound source which vibrates back and forth 1,000 times per second has a frequency of 1,000 cycles per second (cps). Frequency is now stated in hertz (Hz) instead of cps (named after Heinrich Hertz, a German pioneer in research on the transmission of radio waves).

Frequency response—the range of frequencies that an audio device will reproduce at an equal level (within a tolerance, such as 3.0 dB). It is a way of understanding how a microphone responds to sound sources and is usually expressed in graph form, where the horizontal axis represents frequency and the vertical axis amplitude (in dB).

Fundamental—the lowest frequency in a complex waveform. The fundamental is perceived as the pitch of a note.

Harmonics—*see* Overtone series.

Headroom—the difference between the average or nominal level of a signal (in EBU terms, this is the target loudness) and the point at which the signal clips (0.0 dBFS in digital systems).

Hertz (Hz)—the term used to designate frequency in cycles per second and named after the German physicist Heinrich Hertz. It was adopted as the international standard in 1948.

Impulse response (IR)—the reverberation characteristics of an ambient space. An IR is recorded using a short burst of sound (for example, a starter pistol) or a full-range frequency sweep played through loudspeakers to excite the air molecules in a room. After the sound of the stimulus has been removed from the recording (through a process known as deconvolution), the room's impulse response or reverb tail can be added to a dry signal.

Inverse Square Law—in the direct or free field (that is, in a field free from reflections), soundwaves radiate in all directions from a source in ever-expanding spheres, and as the surface areas of these spheres increase over distance, the intensity of the sound decreases in relation to the area the soundwaves spread across (see Figure Glossary.4). The Inverse Square Law states that the intensity of a sound decreases proportionally to the square of the distance from the source. In other words, for every doubling of the distance, the sound pressure reduces by half, which the human ear perceives as a decrease of 6.0 dB (note that this principle applies only in the direct or free field; in enclosed spaces, the actual decrease is somewhat less than 6.0 dB).

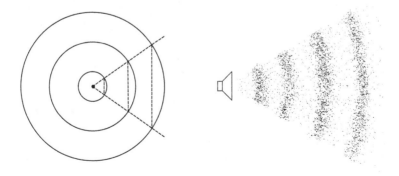

Figure Glossary.4 Spherical propagation of soundwaves in an open space.

K-weighting—a filter that approximates human hearing by de-emphasizing low frequencies (to make them less loud) and emphasizing higher frequencies (to make them louder) (see Figure Glossary.5).

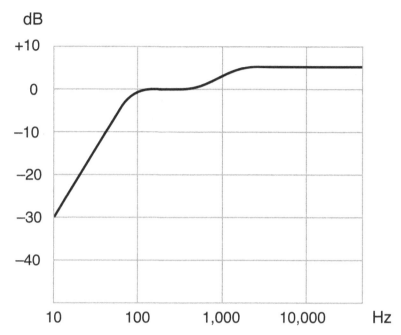

Figure Glossary.5 K-weighting.

Late reflections—*see* Diffuse or reverberant sound field.

Line level—refers to the average voltage level of an audio signal. In professional signal processing components, it is usually +4.0 dBu (dBu is the signal level expressed in decibels referenced to voltage).

LKFS (Loudness, K-weighted, referenced to digital Full Scale)—loudness level on an absolute digital scale. It is analogous to dBFS, for one unit of LKFS equals one dB. This terminology is used by the International Telecommunication Union and the Advanced Television Systems Committee (USA); it is identical to LUFS.

Logarithmic—instead of dealing with a number itself, the number is represented by its logarithm (often abbreviated as log). The common log of a number is the power to which the number 10 must be raised to obtain that number; for example, 10 to the power of 2 (10^2) equals 100, thus the log of 100 is 2. In a logarithmic scale, distances are proportional to the logs of the numbers represented, but in a linear scale the distances are proportional to the numbers themselves.

Lossless—a codec for reducing the size of a file that preserves the original data during coding and decoding; that is, no information is lost in the process.

Lossy—a codec that removes information from an audio signal in order to reduce the size of the file. Principles of psychoacoustics are used to identify parts of the signal that humans cannot hear well, and the codec discards less audible components, which has a detrimental effect on sound quality.

Loudness—a perceptual quantity: the magnitude of the physiological effect produced when a sound stimulates the ear. This physiological reaction is measured by meters employing an algorithm developed by the International Telecommunication Union (ITU) designed to approximate the human perception of level.

LRA (Loudness Range)—originally developed by TC Electronics, it is the overall range of the material from the softest part of a signal to the loudest part, given in LU. To avoid extreme events from affecting the reading, the top 5% and the lowest 10% of the total loudness range is excluded from the measurement (for example, a single gunshot or a long passage of silence in a movie would result in a loudness range that is far too broad).

LU (Loudness Unit)—a relative unit of loudness referenced to something other than digital full scale. It employs K-weighting and is analogous to dB, for one LU equals one dB. This terminology was established by the International Telecommunication Union (ITU).

LUFS (Loudness Unit, referenced to digital Full Scale)—loudness level on an absolute digital scale. It is analogous to dBFS, for one unit of LUFS equals one dB. LUFS employs K-weighting and is identical to LKFS. This terminology is used by the European Broadcasting Union (EBU).

Maximum sound pressure level—the maximum sound pressure level a microphone will accept, while producing harmonic distortion of 0.5% at 1,000 Hz.

Near field—the sound field immediately adjacent to a source where direct sound energy dominates. For microphones, this is the distance within which reflected sound remains minimal.

Noise floor (self-noise)—the internal noise level generated by a device or system (for example, a microphone in the absence of soundwaves striking the diaphragm). The noise comes from the resistance of the coil or ribbon in electromagnetic mics and from the thermal noise of the resistors, as well as the electrical noise of the pre-amp, in electrostatic mics. It is expressed in dB (lower numbers are better).

Normalization—a method of adjusting loudness so that listening levels are more consistent for audiences.

Nyquist Theory—between 1924 and 1928, Harry Nyquist discovered that an analog signal can be recreated accurately only if measurements are taken at a rate equal to or greater than twice the highest frequency in the signal. The maximum frequency a digital system can represent is about half the sampling rate.

Overtone series—the frequencies above the fundamental in a complex waveform, that is, a collection of sine waves integer multiples of the fundamental frequency. This series of frequencies gives notes their tonal color or timbre.

"Passive" device—one that does not require external power to function.

Periodic soundwave—a waveform that repeats its shape. All waveforms with pitch are periodic.

Phase—the starting position of a periodic wave in relation to a complete cycle.

Phon—a unit used to relate perceived loudness to the actual sound pressure level of a signal. Phons describe the psychological effect of loudness. The concept of the phon is part of the system known as the equal loudness curves or contours, a system in which the threshold of hearing (0.0 dB) for a 1 kHz sine wave (pure tone) is equated to 0 phons.

Plane soundwave—when waves propagating spherically reach the point at which the surfaces of the spheres become almost flat, the intensity of the waves decrease in a linear fashion, more or less uniformly. At this distance, the Inverse Square Law no longer applies, because the total area of the plane changes very little as the waves travel forward.

PLR (Peak-to-loudness ratio)—the difference between a signal's maximum true-peak level and its integrated or average loudness.

Polar coordinate graph—a graphing technique used for plotting the directional sensitivity patterns of microphones. Concentric circles represent the sensitivity in terms of dB, and the plotted lines show the amount of attenuation that occurs for specific frequencies arriving from various angles.

Polar patterns—the polar response of a microphone indicates its sensitivity to sounds arriving from any location around the diaphragm.

Pre-delay—the time gap between the arrival of the first wavefront at a listening position in an enclosed space and the arrival of the first reflection from a nearby surface (also known as the initial-time-delay gap).

Pressure-gradient transducer—a microphone operating on differences in pressure from soundwaves arriving on both sides of a single diaphragm or on the outer surfaces of two diaphragms joined together (but separated by a backplate).

Pressure transducer—microphone designers clamp a single circular diaphragm inside a completely enclosed casing so that only the front face is exposed to the sound field. Sounds arriving from all directions exert equal force on the diaphragm, and because the diaphragm responds identically to every pressure fluctuation on its surface, these microphones exhibit a non-directional, that is, an omnidirectional (360°), response pattern.

Proximity effect—the discernible increase in the low-frequency response of pressure-gradient microphones (cardioids and ribbons) as sound sources move closer to the diaphragm.

Pulse code modulation (PCM)—invented by Alec Reeves in the late 1930s, PCM has become the standard method for digitally encoding analog waveforms (the technique is used in both WAV and AIFF). It has three components: sampling, quantizing, and encoding.

Quantization—when the voltage measurement at a sample falls between two of the integers in a scale based on bit depth, quantization rounds (quantizes) the measurement to the closest step of the scale.

Quantization noise—the rounding of measurements taken in the sampling process introduces errors into the system (heard as nonrandom noise), and the size of the error depends on the number of steps the scale contains: a 2-bit scale has 4 possible steps (2^2), a 3-bit scale 8 steps (2^3), a 4-bit scale 16 steps (2^4), an 8-bit scale 256 steps (2^8), a 16-bit scale 65,536 steps (2^{16}), and a 24-bit scale 16,777,216 steps (2^{24}). Scales based on higher numbers of bits, then, because they have more finely graded steps, reduce the size of the rounding error and, hence, the amount of noise in the system. In both 16 and 24 bit scales, the rounding error is so small that the noise introduced by quantization is quite faint.

Resolution—an indication of the sound quality of digital audio based on sample rate and bit depth. Today "high-resolution audio" has a bit depth of at least 24 and a sample rate at or greater than 88.2 or 96 kHz. The greater the "resolution" of the system, the more accurately it can represent waveforms.

Reverberation—*see* Diffuse or reverberant sound field.

Reverberation radius—*see* Critical distance.

Reverberation time (RT)—after a sound source has stopped emitting soundwaves, the time required for the reverberant field to decrease to one-millionth of its original strength, a reduction of 60.0 dB.

Ribbon microphone—these microphones operate on the principle of electromagnetic induction. A thin strip of corrugated aluminum (the diaphragm) is suspended in a magnetic field so that both sides engage with the sound source. These microphones induce an electrical current proportional to the velocity of displacement.

Sample peak—the peak level of a signal that occurs at sampling points.

Sampling—the process of measuring the voltage of an electrical audio signal at a regular interval so that the measurements can later be outputted as binary numbers.

Self-noise—*see* Noise floor.

Sensitivity—the ratio between the electrical output level of a microphone and the sound pressure level on the diaphragm. Usually expressed in dB, it is a measurement of the output produced when a mic is subjected to a standardized sound pressure level (that is, it indicates how much signal any given SPL produces).

Signal-to-noise ratio (SNR)—the ratio between the useful signal produced by a device and its inherent noise when the signal is removed, expressed in dB (higher numbers are better). A signal-to-noise ratio of 47.0 dB means that the noise floor is 47.0 dB below the signal.

Sine wave—a periodic waveform consisting of a single frequency. A sine wave has pitch but lacks the timbral quality associated with the complex waveforms produced by musical instruments and voices.

Sound pressure level (SPL)—soundwaves cause the air pressure at any given point in a wave's cycle to vary above (compression) or below (rarefaction) barometric pressure. This variation in pressure quantifies the strength of a sound and is called sound pressure (this is what a microphone measures). When expressed on a decibel scale, it is called sound pressure level (20 micropascals of air pressure is 0.0 dB on the scale, and this corresponds to the threshold of hearing at 1,000 Hz for a normal human ear).

Spherical soundwave—close to a small sound source (such as the human voice), waves propagate spherically; that is, they travel away from the source in spheres that continuously increase in diameter. These waves decrease in intensity quite rapidly, falling by 6.0 dB for every doubling of the distance in a field free of reflections (the Inverse Square Law).

Transducer—a device that converts one form of energy to another (verb: transduce).

Transient—any sudden and brief fluctuation in a signal or sound that disturbs its steady-state nature. Transients generally are of a much higher amplitude than the average level and often cause devices to overload. The initial peak in energy at the beginning of a waveform (the "attack") is called an onset transient (examples: a word which starts with a consonant, the hammer of a piano striking the strings, a rim shot on a snare drum).

Transient response—a measure of the ability of a device to handle and faithfully reproduce sudden fluctuations. In microphones, it is a measure of how quickly a diaphragm responds to abrupt changes in sound pressure (lighter diaphragms respond more quickly).

True peak—the undetected peak level of a signal that occurs between sampling points.

Index

AAC (advanced audio codec) 79, 106, 112
AAC iTunes+ 107
A-B spaced microphone technique 54–8, 126,
    127, 128, 129, 130, 131, 135, 136, 137,
    138, 148
Adagio see Barber, S.
ADC (analog-to-digital converter) 13, 15, 16
Addison, J. 147
AES (Audio Engineering Society) 112, 113;
    streaming practice 117, 119
AIFF 14, 105
AKG 138, 140, 148
ALAC (Apple lossless audio codec) 106
Albinoni, T. 143, 146, 147; "Amor, sorte,
    destino," *12 Cantate da camera a voce sola*,
    op. 4 143, 144, 145, 147, 149, 150
algorithm 89, 101, 105, 107, 110, 149
aliasing 15
Alkan, C.-V. 137; *Troisième recueil de chants*, Op.
    65 137
"Amor, sorte, destino" *see* Albinoni, T.
analog signal 13, 14, 15, 18, 60
An die Musik 139
angle of acceptance (microphones) 28, 46, 54
angle of incidence (soundwaves) 26, 32,
    52, 53
Apple Music 143
aria 144, 146, 149
Ashbourne, K. 143, 149
ATSC (Advanced Television Systems
    Committee) 107, 110; A/85 117
audio resolution 18–19
audio volume 107
Avison, C. 146

Bach, J. S. 138
backplate 24, 25
baffle 139, 150
Barber, S. 115; *Adagio* 115
BBC iPlayer Radio 117
bell filter 83, 99, 102; sweeping with 70
Benchmark Media 18

Beranek, L. 10, 11
Beyerdynamic 139
bi-directional/figure 8 polar pattern 23, 25, 26,
    28, 29, 33, 35, 48, 49, 56, 58, 62, 128, 138
binary word 15, 17, 18
binaural hearing 53
bit 14, 15, 16, 17, 18
bit depth 14–15, 17, 18, 19, 96, 97, 106
bit rate 15, 105, 106, 110–11
bleed 129
Blumlein, A. 48
Blumlein coincident microphone technique
    48–9, 128, 138
bps (bits per second) 15
Brainworx Audio 82, 83, 150; *bx_dynEQ V2*
    82, 150
brickwall limiter 76, 79
broadcasting 107, 108, 112, 113
bussing 150
byte 14–15

Caccini, G. 151n2
Camerer, F. 113
cantata 143, 147, 148
capacitor 24, 28
cardioid polar pattern 26, 27, 28, 29, 33, 34,
    35, 47, 49, 52, 53, 57, 58, 62, 126, 128,
    130, 131
Case, A. 89
CD 15, 111, 113
CD Baby 143
cello *see* microphone placement
chest voice 144
choir (chamber) *see* microphone placement
chorus 102
"Claire de lune" (*Night Songs*) 115, 116
clarinet *see* microphone placement
Clement, F. 144
clip gain 149
clipping 60, 61, 84, 100, 109, 110, 112, 119
codec 79, 105, 106, 110, 111, 112, 119, 150
coincident microphone techniques 47–51

comb filtering 39, 40, 51, 52, 56, 126, 129, 131

complex waveform 5, 6, 37, 39

compression 73–9, 84, 86, 96, 113, 115, 149, 150

compressor 73–9, 81, 84, 96

concert hall 10–11, 42, 45, 56, 62, 99, 129, 150

condenser microphone 24–28, 132

constructive interference 38, 39

container 105

convolution reverb 102–3, 143

Corri, D. 146, 147

Crane Song 150; *Phoenix II* 150

critical distance 35, 54, 61–2, 129, 130

critical listening 59–60

cross-over frequency (reverb) 92, 97, 102

cut-off frequency (EQ) 65, 94, 99

DAB+ Radio 117

DAC (digital-to-analog converter) 14, 18

damping (reverb) 92–3, 99

Danish State Radio 49

DAW (digital audio workstation) 13–14, 59

dBFS 109

dBTP 110

decay time *see* reverb time

Decca Records 57

Decca tree spaced microphone technique 57–8, 136, 138

deconvolution 102

decorrelation 18

de-esser 84–8

delay 8, 27, 39, 40, 43, 45, 55, 91, 94

depth perspective 9, 59, 62, 135, 136

Desert Fish Studios (Toronto) 139

destructive interference 38, 39

diaphragm 13, 23, 24, 25, 26, 27, 28, 29, 36, 37, 38, 47, 125, 130, 131, 132

diffuse field 25, 31, 34, 35, 61, 62

DIN (Deutsches Institut für Normung) 53

DIN near-coincident microphone technique 53

direct field *see* free field

direct sound 8, 9, 13, 25, 28, 31, 33, 34, 35, 49, 54, 59, 61, 62, 89, 90, 97, 99, 126, 127, 129, 130, 138, 139, 151n1

directivity factor *see* random energy efficiency

distance factor 34–5, 62

distortion *see* noise and distortion

dither 17, 18, 82; *see also* noise shaping; TPDF

Dowland, J. 146

DPA 148, 150

dry sound *see* direct sound

dynamic EQ 82–4

dynamic microphone 28–9

dynamic range 15, 17, 18, 19, 59, 60, 73, 96, 115, 148, 150

Eadon, S. 137

early music 143

early reflections 8, 9, 11, 12, 89, 94, 95, 97–8, 99, 101, 103, 126, 129, 139, 150, 151n1

EastWest 102, 103; *Spaces II* 102–3

EBU (European Broadcasting Union) 59, 107, 108, 110, 111, 112, 113, 115, 116, 117; R 128 117

editing 14, 73

electromagnetic 28, 29

electrostatic 24

EMAC Recording Studios (London, Canada) 143, 148

emphasis (in speaking and singing) 144, 145, 146, 147

EQ 65–71, 84, 89, 91, 92, 94, 99, 101, 102, 149; *see also* filter

equal loudness curves (contours) 71

European television standard of loudness 117

Eventide Audio 149; *UltraReverb* 149

Exponential Audio 92, 101; *Nimbus* 92–7

FabFilter 68, 101; *Pro-Q 2* 68–9; *Pro-R* 101–2

"Fantasie" (*Secret Fires of Love*) 117, 119

Faulkner phased-array microphone technique 56–7, 139–40

Faulkner, T. 56

Ffrangcon-Davies, D. 145

FFT (fast Fourier transform) 83, 88

figure 8 *see* polar pattern

file size 106–7

filter, definition of 65

filter slope 65–6

FLAC (free lossless audio codec) 105–6, 107

Fleming, R. 115; *Dark Hope* 115–16; *Night Songs* 115–16

Fletcher, H. 71

flute *see* microphone placement

Flux 97, 101; *Verb Session* 97–100

Fraunhofer Institute for Integrated Circuits 106

free field 8, 9, 25, 31, 33, 35
French horn *see* microphone placement
frequency *see* soundwave
frequency band 19, 66, 68, 69, 82, 83, 84, 86, 87, 88
frequency response 25, 31, 41
fundamental 5, 6, 126, 128

García, M. 146
graphic filter 69
Großer Musikvereinssaal (Vienna) 98
Guarnerius, J. 138
GUI 80, 83, 92, 97, 103
guitar 143

Hamelin, M. A. 137
Handel, G. F. 115; "He was despised" (*Messiah*) 115
harmonic series 5, 6, 59, 60, 126, 128
harpsichord 143, 148, 149, 150
headphone 53, 83
headroom 60, 111, 118, 119
head voice 144
Henry Wood Hall, London, UK 137
"He was despised" *see* Handel, G.F.
high-resolution audio 19
historically informed performance 143, 144–8, 149, 150
hypercardioid polar pattern 27, 33, 34
Hyperion 137

"I attempt from love's sickness to fly" *see* Purcell, H.
impulse response (convolution reverb) 89, 102, 143
Institute of Sound Recording, University of Surrey 138
integrated (average) loudness 108, 112, 113–15, 117, 118, 119, 121n2 (definition)
inter-sample peak level *see* true-peak level
Inverse Square Law 8, 35, 36, 61, 149
iPad 106
iPhone 106
iPod 106
Isabel Bader Centre for the Performing Arts, Queen's University (Canada) 139
isolation booth 148
ITU (International Telecommunication Union) 107, 108, 110, 113; BS.1770 117

iTunes 106; *Sound Check* 117
iZotope 112, 149; *Rx 6 Advanced* 112, 149

Jalpert, D. 139
Jecklin Disk *see* OSS near-coincident microphone technique
Jecklin, J. 53

Katz, B. 109
K-weighting (loudness) 109, 110

late reflections 8, 9, 11, 89, 90–1, 93, 94, 95, 98, 99, 101, 102
Lauridsen, H. 49
Leininger, T. 143, 148, 149–50
level, setting of 60–1
Lexicon 480L 138
limiter 76, 79–82, 84, 96, 119; *see also* brickwall limiter
LKFS 110
lossless compression 105–6
lossy compression 105, 106, 110, 112, 119
loudness 59, 71, 73, 74, 77, 90, 107, 108, 150
LRA (loudness range) 108, 110, 115, 117, 118
LU (loudness unit) 110
LUFS 110
lute 143

m4a 106
m4p 106
Mancini, G. 144
Mass see Vaughan Williams, *Mass*
messa di voce 144, 146, 149
McDSP 150; *FilterBank F202* 150
metadata 105
meter 107, 108, 109; absolute scale in 109; relative scale in 109; EBU mode of 108; sample peak in 110; true peak in 110–11
MeterPlugs 116; *LCAST* 116–18
Milab 148
microphone placement: cello 131, 138; chamber choir 136; clarinet 131; double bass and piano 139–40; flute 132; French horn 132–3; oboe 131; piano (solo) 125–8, 137, 138; piano trio 135; string quartet 135–6; trombone 132; trumpet 132; violin and piano 130–1; voice 129–30
mixing 14, 59, 60, 71, 73, 89, 108, 113, 119, 136, 143, 148, 149, 150

momentary loudness 108, 117

monophonic broadcasting 52

Monteverdi, G. 107; "Sì dolce è'l tormento" 107

moving coil microphone *see* dynamic microphone

mp3 79, 106, 107, 110, 112

MPEG (Moving Picture Expert Group) 106

M/S (mid-side) coincident microphone technique 49–51, 58, 127

Munson, W. 71

Nation, R. 143, 148, 149–50

near-coincident microphone technique 51–4

near field 36, 61, 84, 125

Neumann 139

noise and distortion 59, 60, 79, 96, 106, 112, 113, 119, 149

noise floor 18, 60

noise shaping 17, 18, 82; *see also* dither; TPDF

noise types: blue 17; nonrandom 16, 17; random 17; shaped 17, 18; white 17

normalization 108, 109, 110, 113, 118–19

normalization target 109, 112–13, 118–19

NOS (Nederlandse Omroep Stichting) 52

NOS near-coincident microphone technique 52

notation, interpretation of 146, 147

notch filter 86, 102

Nugen Audio 18, 79, 116, 117, 150; *ISL 2* 79–82, 150; *MasterCheck* 117–20, 150; *SEQ-ST* 150; *Stereoizer* 150

null areas 24, 33, 49, 131, 138, 139

Nyquist frequency 15

Nyquist, H. 15

oboe *see* microphone placement

Office de la Radiodiffusion-Télévision Française 51

Ogg container 105, 106, 107, 112

omnidirectional polar pattern 24, 25, 26, 28, 31, 32, 34, 48, 53, 54, 56, 57, 62, 126, 127, 128, 129, 130, 131, 135, 137, 138, 148

onset 77

opera house 10–11

oration 144, 145, 147

orchestral hall 103

Ornithoparchus, A. 146

ORTF near-coincident microphone technique 51–2, 57, 126–7, 128, 136, 139, 148

OSS (optimum stereo signal) near-coincident microphone technique 53–4, 148

oversampling 110

overtones *see* harmonic series

Pandora 117

parametric filter 68, 82, 83, 102, 150

partials *see* harmonic series

pass filter 65–7, 68, 70, 82, 86, 86, 87, 88, 95, 103, 149, 150

pauses (in speaking and singing) 144, 145, 146, 147

PCM (pulse code modulation) 14, 15, 105, 110

peak level 60, 73, 74, 79, 82, 83, 87, 88 107, 109, 149

perceptual coding 106

period *see* soundwave

phantom image 45, 47, 49, 55

phantom power 24

phase 26, 37–42, 45, 48, 49, 51, 52, 56, 58, 91, 126, 131, 138

phrasing 144, 146

piano: location in the recording space 125–6; soundwave propagation from 125; stereo mic placements 126–8, 137, 138; unfavorable room acoustics 127–8

piano trio *see* microphone placement

pick-up pattern *see* polar pattern

plate reverb 94

PLR (peak-to-loudness ratio) 117, 118–19

plosive 130, 149

polar pattern 31–3

Pong Studio (Toronto) 138

pop screen 132, 148

popular music vs. classical music 115–16

portamento 144, 146

pre-amp 24, 30

pre-delay (reverb) 10, 11, 89–90, 92, 93, 98, 99, 101, 103

presence 11, 99, 101, 129, 130, 131, 135

pressure microphone 24, 25

pressure-gradient microphone 23, 25–8, 35

propagation of soundwaves from instruments and voices: clarinet 131; flute 132; oboe 131; piano 125–6; trombone 132; trumpet 132; violin 130; voice 129, 149

prosody 144, 145, 146, 147, 149

Pro Tools 148, 149

proximity effect 35–7, 126, 130, 131, 148, 150

psychoacoustics 17, 45, 105, 106, 107, 109
pumping (compressor) 77, 113
punch-in 147, 149
punctuation 144, 145
Purcell, H. 111; "I attempt from love's sickness to fly" 111

Q (quality factor) in filtering 68, 69, 83
quantization 14, 16, 17, 18
quantization error 16–17
Quarrington, J. 139

Ravel, M. 138; *Sonatine, No. 2* 138
recitative 144, 146, 149
reconstruction filter 18
re-creative performance 146
Red Maple Sound (Toronto) 138
REE (random energy efficiency) 33–4
Reeves, A. 14
reflection simulation 89–102, 143, 150
register (voice) 144–5; *see also* chest voice; head voice
Replay Gain 117
requantization 17
reverb tail *see* late reflections
reverb time (decay rate) 8, 10, 12, 54, 60, 89, 90–1, 92, 98, 99, 101, 102, 103, 151n1
reverberant field *see* diffuse field
reverberation, definition of 7–8
reverberation radius *see* critical distance
rhetoric 144, 145
ribbon microphone 28–30, 132, 138, 139, 148, 149
RIFF container 105
RMS (root mean square) 73
room size (reverb) 92, 98–9, 151n1
Royer 138, 148, 149, 150
rubato 144
"Ruhe, meine Seele!" (*Night Songs*) 115

sample (sampling) 14, 15, 16, 17, 43, 102, 107, 110
sample rate 15, 18, 19, 106, 111, 112,
sample rate conversion 106, 111, 112, 150
saturation 97
Schoeps 137, 138, 148
Searles, R. 138
Secret Fires of Love 107, 111, 117, 143, 144, 147, 150

shelf filter 67, 68, 69, 70, 82, 83, 99, 102
short-term loudness 108, 117, 118
Siau, J. 18
"Sì dolce è'l tormento" *see* Monteverdi, G.
sibilance 84, 86, 87, 130, 148
signal flow 13–14
sine wave 3, 4, 5, 38, 39
SNR (signal-to-noise ratio) 18, 60, 148
Sonatine, No. 2 see Ravel, M.
Sonnox 78, 83, 86, 89, 101, 107, 112; *Fraunhofer Pro-Codec* 107, 112; *Oxford Dynamic EQ* 83–4; *Oxford Dynamics* 78; *Oxford Inflator* 150; *Oxford Reverb* 89–91; *SuprEsser* 86–8
soundboard 125
sound intensity 35, 45, 51
sound pressure level 35, 36, 61–2
Sound Radix 43; *Auto Align* 43
soundwaves: compression 3, 13; cycle 3, 4; frequency 4; period 3, 4, 5, 37; rarefaction 3, 13
spatial environment 59
SPL (sound pressure level) meter 61
Spotify 117, 119, 143
spot mic 129, 130, 131, 135, 136, 140
Steinberg 112; *Wavelab Pro 9.5* 112
stereo image 45, 47, 48, 49, 50, 51, 52, 53, 54, 55, 56, 58, 59, 60, 91, 125, 126, 127, 130, 138
stereo perception, optimum angle 45
stereo sound stage 45, 48, 49, 50, 54, 55, 56, 91, 130
streaming 60, 73, 105, 112, 113, 117, 119, 143
string quartet *see* microphone placement
Studer A80 138
Studio Rhetorica 107, 117, 143
studio techniques 143–51
supercardioid polar pattern 27, 33, 34

TC Electronics 110
tempo 144, 145, 146–7
texture 89
Thomson, D. 143, 144, 145–6, 147, 148, 149–50
three-to-one principle 39–43, 131
Tidal 117
timbre 5, 6, 28, 59, 60, 89, 99, 125, 126, 144–5, 148
time of arrival 45, 51, 53, 55, 56

"Today" (*Dark Hope*) 115, 116
tone color *see* timbre
TPDF 17, 18, 82; *see also* dither; noise shaping
transducer *see* microphone
transduction 23
transient 60, 84, 87
transparency 59
transistor 24
Troisième recueil de chants, Op. 65 *see* Alkan, C-V.
trombone *see* microphone placement
true-peak level 60, 79, 107, 108, 110, 111, 115, 117, 118, 119, 150
trumpet *see* microphone placement
truncation 17

unfavorable room acoustics *see* piano
Universal Audio 149; *EL7 Fatso Sr* 150; *Lexicon 224 Digital Reverb* 149–50; *Massenburg DesignWorks MDWEQ5* 149, 150

vacuum tube 24
Vaughan Williams, R. 115
Vaughan Williams, *Mass* 115

Vicentino, N. 144, 146–7
violin *see* microphone placement
voice *see* microphone placement
voltage 13, 14, 15, 16, 17, 18, 24
Vorbis 106, 107, 112
Voxengo 69
Voxengo, *Marvel GEQ* 69

Wallace, R. 57, 58
WAV 14, 105, 107
Weiss Engineering 107, 112; *Saracon* 107, 112
"With twilight as my guide" (*Dark Hope*) 115
Wolpert, J. 139, 140
Woram, J. 36
word length *see* bit depth
wrapper 105

Xiph.Org Foundation 105, 106
X-Y coincident pair microphone technique 47–8, 58, 128

YouTube 117

Zelenka, W. 138